THERMAL ENERGY STORAGE IN AQUIFERS

Pergamon Titles of Related Interest

Andersen & Moreland   Hydrides for Energy Storage
Blair   Aspects of Energy Conversion
Kovach   Thermal Energy Storage
MacKenzie-Kennedy   District Heating
Reay   Industrial Energy Conservation, Second Edition

*Related Journals**

Energy
Energy Conversion and Management
International Journal for Housing Science and its Applications
International Journal of Multiphase Flow
Physiochemical Hydrodynamics, The International Journal
Progress in Surface Science
Solar Energy
Sun World
Underground Space

**Specimen copies available upon request.*

Thermal Energy Storage in Aquifers

Design and Applications

Walter J. Schaetzle
C. Everett Brett
David M. Grubbs
Marvin S. Seppanen

Pergamon Press

New York • Oxford • Toronto • Sydney • Paris • Frankfurt

Pergamon Press Offices:

U.S.A. Pergamon Press Inc., Maxwell House, Fairview Park,
Elmsford, New York 10523, U.S.A.

U.K. Pergamon Press Ltd., Headington Hill Hall,
Oxford OX3 0BW, England

CANADA Pergamon of Canada, Ltd., Suite 104, 150 Consumers Road,
Willowdale, Ontario M2J 1P9, Canada

AUSTRALIA Pergamon Press (Aust.) Pty. Ltd., P.O. Box 544,
Potts Point, NSW 2011, Australia

FRANCE Pergamon Press SARL, 24 rue des Ecoles,
75240 Paris, Cedex 05, France

FEDERAL REPUBLIC OF GERMANY Pergamon Press GmbH, Hammerweg 6, Postfach 1305,
6242 Kronberg/Taunus, Federal Republic of Germany

Copyright © 1980 Pergamon Press Inc.

Library of Congress Cataloging in Publication Data

Main entry under title:

Thermal energy storage in aquifers.
  Bibliography: p.
  Includes index.
1. Heat storage. 2. Aquifers. 3. Heating.
I. Schaetzle, Walter J.
TJ260.T48 1980   621.402   80-20239
ISBN 0-08-025977-4

Printed in the United States of America

The Authors

Walter J. Schaetzle, B.S., M.S. and D.Sc. in ME; Professor of Mechanical Engineering, The University of Alabama, Tuscaloosa; President, W. J. Schaetzle & Associates, Inc; formerly with McDonnell-Douglas Corporation; P.E.

C. Everett Brett, Ph.D.; Associate Professor of Geology; Acting Director, School of Mines and Energy Development; Director, Natural Resources Center, The University of Alabama, Tuscaloosa; Vice President, W. J. Schaetzle & Associates, Inc.; formerly with Exxon Corporation.

David M. Grubbs, Ph.D.; Consultant in Natural Resources and Professor Emeritus of Mineral Engineering, The University of Alabama, Tuscaloosa.

Marvin S. Seppanen, B.M.E., M.S. and Ph.D. in I.E.; Associate Professor of Industrial Engineering, The University of Alabama, Tuscaloosa; formerly with the automotive industry and U. S. Bureau of Mines; P.E.

Acknowledgements

The authors gratefully acknowledge the assistance and cooperation of the many persons who contributed to the preparation of this book. Mr. James E. Calm of the Argonne National Laboratory furnished a wealth of data used in the text. Mrs. Jo Ann Owens typed the manuscript with Mrs. Dorthy Schaetzle who prepared equations, tables, figures, and correlated the completion. Additional secretaries in the Department of Mechanical Engineering furnished help on figures and tables. Students at The University of Alabama drafted the majority of the the illustrations. Mr. Jerry LeCroy and Dr. Jasem Ansari, graduate students, compiled performance and aquifer data.

Appreciation is expressed by the authors to the School of Mines and Energy Development and the College of Engineering, The University of Alabama, for their support and assistance in preparing the manuscript for the book.

Mini-Computer Programs

Most calculations in this book are available from the authors as programs for small hand-held computers.

CONTENTS

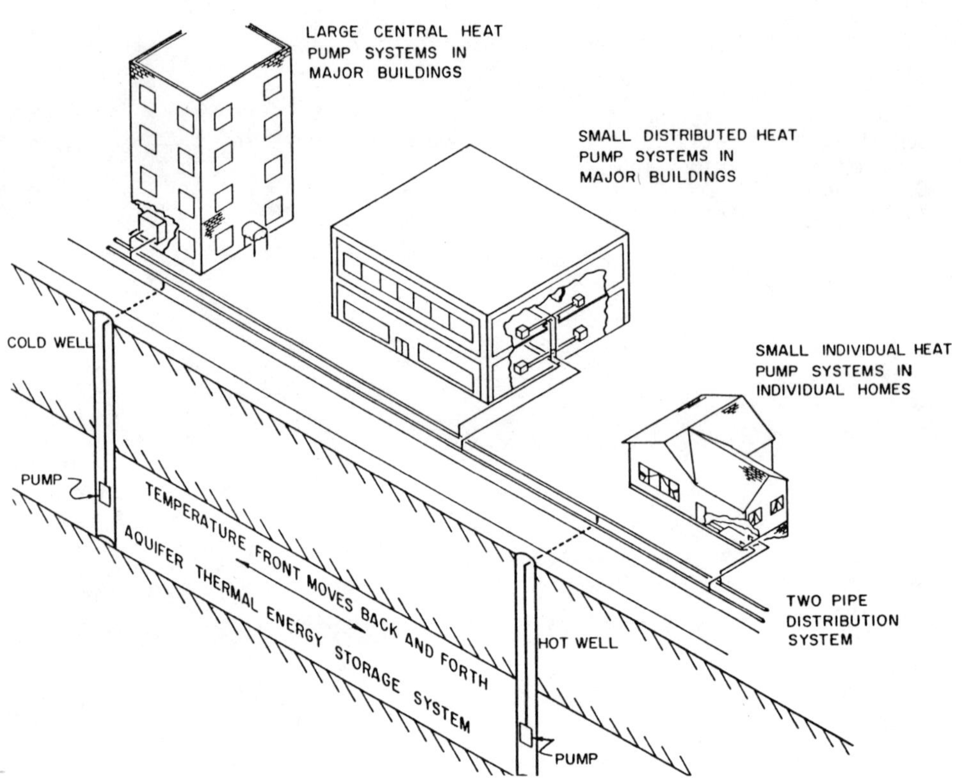

LARGE CENTRAL HEAT PUMP SYSTEMS IN MAJOR BUILDINGS
SMALL DISTRIBUTED HEAT PUMP SYSTEMS IN MAJOR BUILDINGS
SMALL INDIVIDUAL HEAT PUMP SYSTEMS IN INDIVIDUAL HOMES
COLD WELL
PUMP
TEMPERATURE FRONT MOVES BACK AND FORTH
AQUIFER THERMAL ENERGY STORAGE SYSTEM
HOT WELL
TWO PIPE DISTRIBUTION SYSTEM
PUMP

CHAPTER I

INTRODUCTION TO AQUIFER ENERGY STORAGE

A new strategy for storage of energy in naturally-occurring underground aquifers provides the potential for immense conservation of energy and the realization of equally enormous savings in economic costs. Storage of energy in aquifers for subsequent retrieval and use in areas where aquifer systems are of adequate capacity is adaptable for both heating and cooling purposes in homes, public buildings, malls, and ultimately entire cities. The term aquifer is derived from the Latin _aqua_ + _ferre_ which means water-bearing, and more specifically relates to a water-bearing bed or stratum of sand, gravel, or porous rock found near or at nominal depths below the soil zone of the Earth's surface. Water contained in aquifers is referred to as groundwater.

The source of groundwater is frequently thought of as underground streams. Although it is true that, in certain geologic settings characterized by the presence of thick beds of carbonate rock such as limestone, underground streams are found in large solution channels or caves, aquifers are considered to be compact underground rock formations containing groundwater. An aquifer may be thought of as being analogous to a hard sponge.

Aquifers have physical properties described as porosity and permeability, that is, the presence of intergranular voids, fractures, solution channels, or other interconnected openings capable of receiving, storing, and transmitting water from precipitation that infiltrates from the Earth's surface. More commonly aquifers consist of rocks of sedimentary origin such as limestones, dolomites, siltstones, sandstones, and conglomerates although, in some instances, igneous and metamorphic rocks may contain water in sufficient quantities to serve as sources of fresh water.

Approximately 60 percent of the surface of the continental United States is underlain by aquifers adequate for utilization as thermal energy storage systems. A similar percentage would be anticipated in other contentinal areas. During the settling of North America the principal mode of transportation was by water. Therefore, most of the population was concentrated along waterways and, consequently, most of the major urban centers today are located at the same sites. Inasmuch as waterways include both lakes and rivers which are intimately associated with, and in most cases are an integral part of, aquifer systems, aquifers capable of being used for energy storage are available to about 75 percent of the American population.

There are certain inherent advantages in using aquifers for thermal energy storage that should be emphasized. They are: (1) Groundwater may be available within a few hundred feet of the place where it is needed, frequently on the same property, thus avoiding the need for long pipelines and rights-of-way over

2

stretches of several miles. (2) Groundwater may be available for use in areas where other sources of thermal energy storage may be precluded because of economic and environmental considerations. (3) The volume of groundwater in aquifers generally fluctuates little even though the groundwater may be used for consumptive purposes; the water level is not strongly affected by wet and dry periods. (4) Groundwater is uniform in temperature (and in soluble mineral load) throughout the year, and is generally free of turbity and pollution by pathogenic organisms. (5) There is little or no evaporation loss from underground reservoirs, so the risk of depleting the reservoir after a storage system is installed is negligible. (6) Aquifers are ready-made; there are no construction costs other than the installation of the well system for access to the aquifer. (7) Aquifers store thermal energy in both the rock matrix and the contained water, thereby providing a storage capacity of great magnitude independent of surface considerations. (8) Aquifers do not wear out and require little or no maintenance. Life expectancy of a storage system may be infinite.

The Earth is an excellent insulator; this means that thermal energy associated with both hot and cold water injected into an aquifer can be retrieved at near input temperature. Although thermal energy recovery is site-specific, recoveries in the range of 80 percent to 90 percent may be expected.

Access to the energy storage system is by (1) a single well or (2) pairs of wells (Fig. 1.1.) For the single-well system an external source of water is heated or cooled and injected into the well. To recover the energy, the heated or cooled water is withdrawn from the well, is used, and is discarded. For the well-pair system, water is withdrawn from one well, is heated or cooled, and is injected into the second well. To recover the energy the process is reversed. The hot or cold water is withdrawn from the original injection well. After the water is used for heating or cooling, it is reinjected into the original withdrawal well. For every gallon of water withdrawn a gallon of water is reinjected. As a result the net consumption of water at any time is zero. A temperature front moves in the aquifer between the wells as a result of the water flow, and reverses direction seasonally.

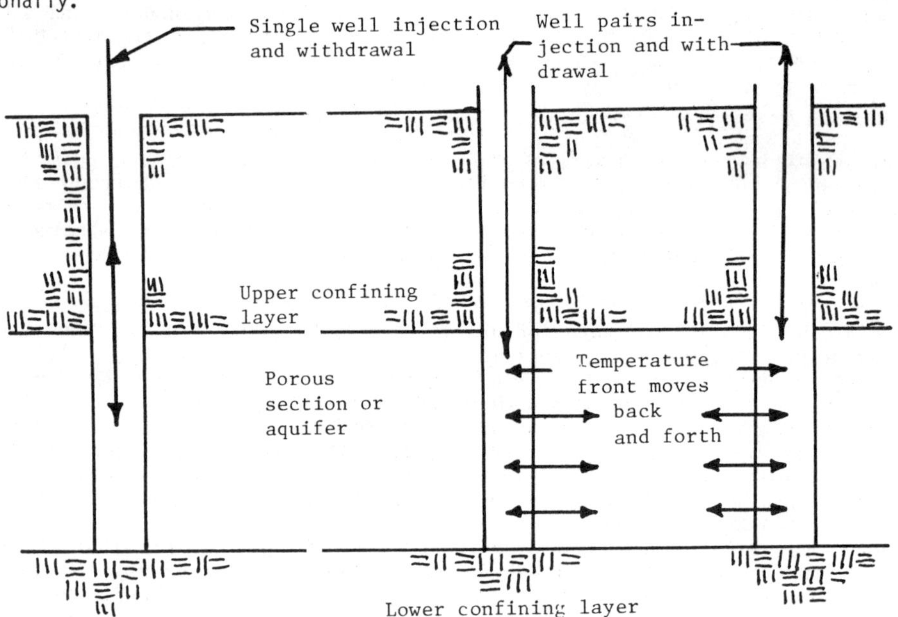

Fig. 1.1 Schematic of single well and well-pair systems.

The storage system is insulated by the confining layers of rock above and below the aquifer. Confining rocks are excellent insulators and may be relatively thick, several hundreds of feet in some cases. In a student project at The University of Alabama, the annual temperature variation below the surface of the Earth was measured. At eight feet below the surface, a very small change, less than one degree F, could be detected. At sixteen feet below the surface, no change in temperature could be detected. A large thermal energy storage system, for example 150 ft thick and 200 ft in diameter, or 4.7×10^6 ft^3, and blanketed by some 60 ft of overburden, provides a large, extremely well insulated thermal energy storage volume, with a small surface area-to-volume ratio. Energy recovery in such a system will exceed 80 percent on an annual injection and withdrawal basis. On a weekly storage basis, recovery will exceed 90 percent, and on a daily basis may exceed 99 percent. Thermal energy losses are residual losses to the rock and water and to the natural flow of water in the system.

Thermal energy storage capacity is a function of the rock, porosity, and temperature variation. An average storage number is 30 Btu/ft^3°F. In the case of the example described in the previous paragraph, the approximate thermal energy storage capacity is 1.4×10^8 Btu per degree temperature change. The storage capacity for a 10°F temperature change is 1.4×10^9 Btu and for a 100°F temperature change is 1.4×10^{10} Btu. Storage capacities a few orders of magnitude higher are a real possibility.

The aquifer thermal energy storage system must be a large system. A 150-ft thick aquifer with a 5-ft diameter core for storage has a volume of 3,000 ft^3 and a storage capacity of 9×10^6 Btu for a 100°F temperature difference. The natural groundwater flow in the aquifer could dissipate this energy prior to the time for its recovery. Large energy storage systems are needed to meet technical requirements of the overall system. The cost of a thermal energy storage system is based on a minimum of two wells and two pumping systems. Thus, the cost is identical for a small storage system (10^6 Btu) or a large system (10^{10} Btu). Economic considerations also demand a large storage system. For a large system the cost per-unit thermal energy is but a small fraction of a man-made system, including man-made lakes. The system is compatible with community heating and cooling systems.

Aquifer Technical Analysis

Research on energy storage in aquifers has provided few experimental results. However, aquifers are beginning to be analyzed for major energy storage projects. The report of the Thermal Energy Storage in Aquifer Workshop (1) includes 87 references on hot and cold water storage in aquifers as compiled by Marcelo J. Lippman. More than a dozen papers by Meyers starting in the 60's (for example Ref. 2, 3, 4) deal with the heat-storage well concept and large storage systems for power plant waste heat. Aquifer study has been worldwide, including work in Switzerland and Germany (5, 6, & 7), Sweden (8), France (9), Japan (1), and the United States (1). Experimental work has been done in France, Germany, the United States, and Japan as abstracted from Reference 10.

France. In 1976 and 1977 the French Bureau of Geological and Mineral Research (BRGM) injected 1400 m^3 of 40°C water into a confined aquifer, three meters thick, at Bonnaud. After a month's storage, 30 percent of the energy was recovered. A more recent and larger scale experiment was conducted at Campuget during 1977 and 1978. Approximately 20,000 m^3 of water at 33.5°C was injected into a shallow, unconfined aquifer and sequentially withdrawn over a five-month period. Energy recoveries were low, mainly due to the shallow storage depth (three to ten meters), significant surface water infiltration, and a two percent hydraulic gradient induced by a pumping well 200 meters away. The results were sufficiently encouraging, however, to motivate a project that would involve the heating of 100 to 500 housing units.

4

Germany. The Geological Survey of North Rhine/Westphalia conducted a thermal storage experiment near Krefeld, Germany in 1973 and 1974. A total of 450 m³ of water at 45°C was injected into a near-surface aquifer. This field experience, along with other studies, prompted the Federal Ministry for Research and Development and the European Community to initiate additional aquifer storage research and development.

United States. At a site near Mobile, Alabama, Auburn University (11) conducted an aquifer storage experiment in a confined aquifer. Two cycles of injection, storage, and recovery were completed. In the first cycle, 55,000 m³ of 55°C water was injected and stored for 48 days, and then withdrawn with a resulting energy recovery of 67 percent. The second cycle was similar, with 58,000 m³ of water being injected at 55°C, and improved energy recovery of 74 percent. Difficulties with chemical-related plugging were encountered in that experiment; however, fluid incompatibility problems may be eliminated through the use of a doublet well scheme where no external water is introduced into the aquifer.

Texas A & M University (12) is conducting a chilled water experiment near College Station, Texas. Between November 1978 and April 1979, approximately 7,500 m³ of 21°C water were pumped from a supply well, chilled to around 10°C in a spray pond, and then injected into a shallow aquifer. Recovery had not been attempted as of late 1979. Significant iron oxide precipitation was observed in the chilled water stream; this is being removed by a sandbed filter.

Japan. In 1977, scientists of the University of Yamagata conducted a heat storage test in a confined shallow aquifer in the Yamagata Basin on the island of Honshu. A well doublet system was used to store heated water in the summer and chilled water in the winter. Heating of the water was accomplished by solar collection and heat exchange with power plant reject water. Chilling was accomplished by use of melting snow on the roof of the research building. Energy recoveries were encouraging (approximately 40 percent), plugging problems were not experienced, and thermal dispersion was at an acceptable level.

Even though thermal storage technology is new, withdrawal of groundwater from aquifers is an old and well-defined practice. Groundwater plays a critical role in the United States with over 50,000,000,000 gallons of water being required daily for public supplies. This includes rural use, irrigation, and industrial use (13). Many technical books are available in the fields of hydrology, porous media, and reservoir engineering.

Numerous studies, both theoretical and applied, are now being conducted. The U.S. Department of Energy through Battelle Northwest Laboratories has awarded a number of two-year design studies for demonstration projects using aquifer thermal energy storage. The U.S. Department of Energy through Argonne National Laboratory is expected to award a number of projects in the heat pump integrated community energy system for final design. A study project (14) by the authors, sponsored by Argonne National Laboratory for DOE, has helped establish a foundation for these projects.

Continuing projects at The University of Alabama, Texas A & M University, and other research centers are establishing a data base. Major and minor computer programs on aquifer thermal energy storage are being developed at the above and other locations. Most aquifer calculations and other calculations in this book are available for Hewlett Packard and Texas Instruments programmable, reading, and writing hand computers from the authors.

The technical capability with supporting demonstrations is developing rapidly for thermal energy storage. This book records a small part of the progress being made. The possibilities of energy savings utilizing the thermal energy aquifer storage system provide a major impetus in this direction.

Potential Application for Thermal Energy Storage

A wide variety of applications are available for large thermal energy storage systems. For many of these systems, energy can be saved with a reduction of capital outlay. Heating and cooling systems operate very efficiently in conjunction with annual thermal energy storage systems. The rejected (heat pump) and collected (solar and waste) energy during warm months can be used for heating during cold months. The solar energy collected during daylight hours in good weather can be used for nighttime and bad weather power generation with a solar power plant. Short discussions of a few of these systems follow. The economical thermal energy storage system provided by nature makes these systems feasible. More detail on these applications is included in later chapters.

Direct (Free) Cooling

In this system in Fig. 1.2 water is cooled in the range of 40°F to 50°F (5°C to 10°C) during cold weather months, stored, and returned for cooling as needed. A reasonable period of sufficiently cold weather is necessary so that large amounts of water can be cooled to the required temperature in either wet or dry cooling towers, a cooling pond, or taken directly from cold streams or lakes. Conditioning the aquifer to a 50°F (10°C) temperature difference, 40°F to 60°F (5°C to 15°C) for example, will minimize heat leaks to the confined aquifer.

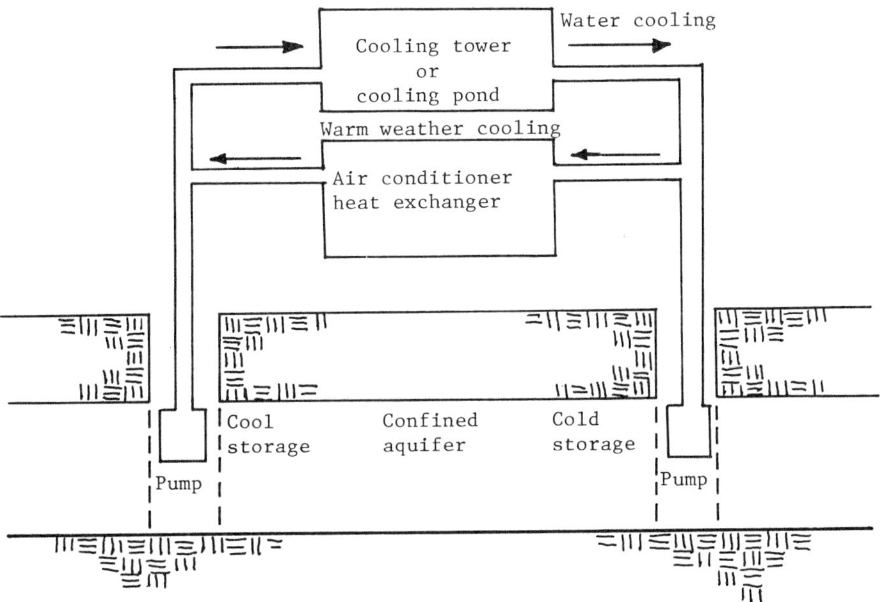

Fig. 1.2. Direct cooling system.

The system in general will have lower capital costs since the refrigeration system is eliminated. The resulting operating cost, a function of energy for pumping requirements, generally will decrease by a factor of well over three to four from the standard air-conditioning system for locations where suitable aquifers are available.

Solar Heating

In this system in Fig. 1.3 solar energy is collected around the year, stored in aquifers, and used as required for heating. By collecting energy on an annual basis, collection takes place at the maximum insolation, warmest ambient temperature, and minimum cloud cover, thereby maximizing collector efficiency. The large storage system eliminates or minimizes the need for a backup system, and reduces the problems which occur with normal zero collection periods and high temperatures which occur in solar collectors (stagnation conditions for zero requirement periods) during the summer. The large size of the aquifer storage system allows lower temperature collecting since energy packing is not critical, again increasing the collector efficiency.

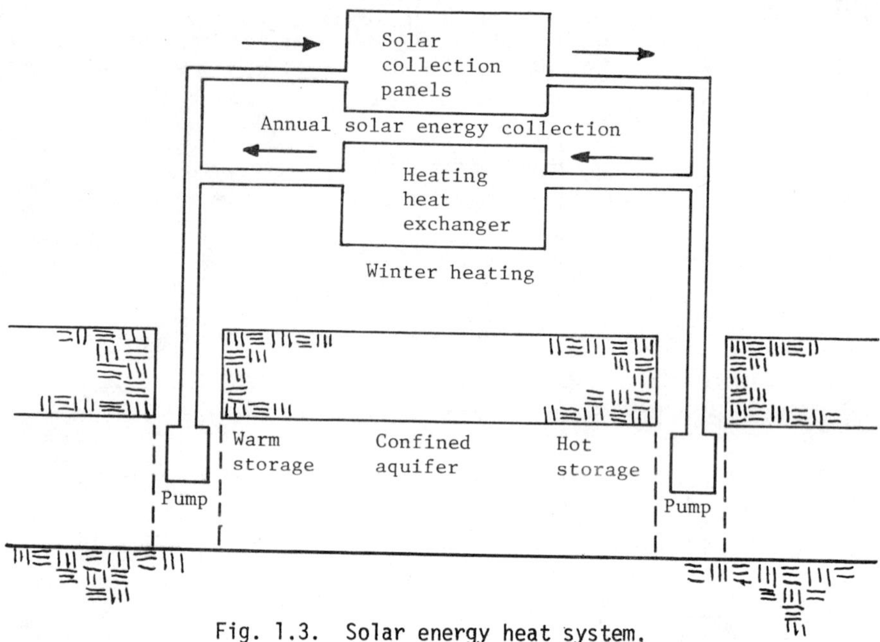

Fig. 1.3. Solar energy heat system.

Combining these factors allows a decrease of 60 to 70 percent in collector area and a corresponding decrease in capital costs. If a suitable shallow aquifer is used an aquifer storage system will have a lower capital cost than the short term storage presently utilized in most solar systems.

Heat Pump Systems

Numerous variations are possible using heat pumps in parallel with an aquifer thermal energy storage system (Fig. 1.4.) Since the aquifer system operates with water, essentially all heat pumps in the system are water source heat pumps. The basic system uses 60°F (15°C) water for cooling during the summer; the heat rejected from the cooling process increases water temperature to 80°F (25°C); the 80°F (25°C) water is stored, retrieved, and used for heating; the heat absorbed for heating reduces the water temperature to 60°F (15°C); and the water at 60°F (15°C) is stored and retrieved for cooling. The energy removed during summer is used for heating and vice versa. With the above temperature range, insulation is not required on the distribution pipelines.

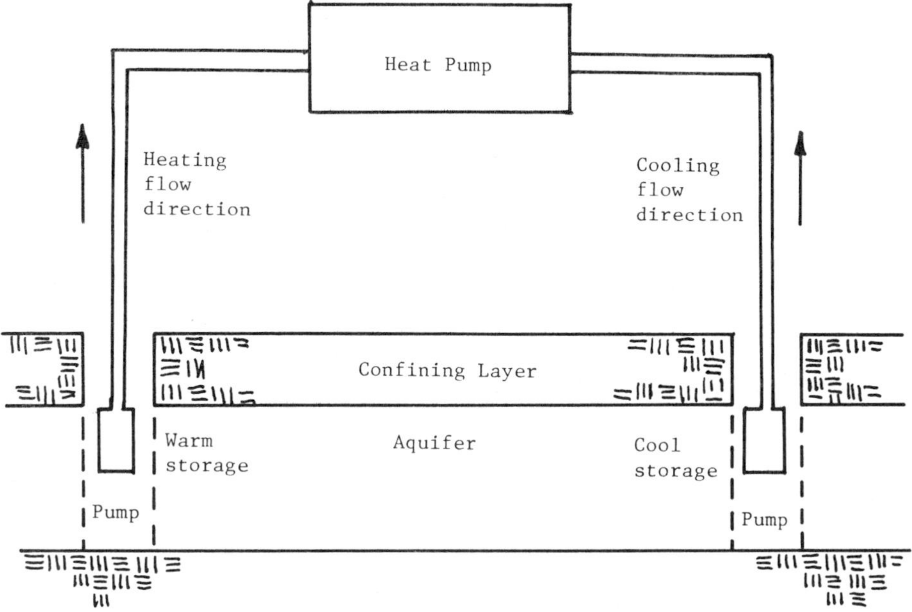

Fig. 1.4. Heat pump system - confining layer.

This system shows a moderate increase in performance for cooling (range of ten to 30 percent) and large increases in performance for heating (30 to 70 percent). The installed capital cost of a water source heat pump is less than that of an air source heat pump system. Cost of an overall heat pump community integrated energy system, including the costs of the wells and the water-distribution system can be more or less than that of the standard system, dependent upon the availability of aquifers and the community density.

By changing temperature ranges, some or all heating or cooling can be by direct heat transfer (8). For example, using 85°F (30°C) water in the heat pump for cooling results in heating the water to 105°F (40°C) with the rejected energy and allows direct heat exchange on the reverse cycle for free heating.

Using 60°F (15°C) water in the heat pump for heating results in cooling the water to 40°F (5°C) due to the absorbed energy and allows direct heat exchange for free cooling. Both these systems (free cooling or heating) in the proper geographical location will reduce energy use by over 50 percent on an annual basis.

Process Steam

By using aquifers with a natural head (pressure) of a few hundred feet of water, storage temperatures of a few hundred degrees C are possible. A system of depressurization is required at the bottom of the well for injected water. Also pressurization is required for recovery to prevent boiling. The water can be either flashed to steam on recovery or used in a boiler to produce steam. Preliminary calculations have shown the storage system cost for daily, weekly, and annual storage for a solar power plant operating at $392°F$ $(200°C)$ (15) is less than one percent of total plant capital cost. As temperatures increase, the work (energy) required to inject and retrieve water becomes excessive.

Other Applications

Numerous other useful applications are available for aquifer thermal energy storage which save energy. Some of these include solar green houses, district heating, hot water applications, etc. Many of the systems provide savings of energy, energy cost, and capital cost. More details of various applications are included in later chapters.

Environmental Considerations in Aquifer Systems

Environmental effects are both adverse and benefical. These include physical, biological, socio-economic, and human environmental changes.

Physical changes in the aquifer include the movement of silt, the dissolving of salts, and change in temperature. These changes are expected to be minor unless outside contamination is introduced into the aquifer. For most heating and cooling systems no major problems are anticipated, especially if the water is isolated from air interfaces (14).

Many types of bacteria can exist in the aquifers if oxygen is available. Iron bacteria can clog aquifer passageways and sand screens. Anaerobes in the form of slime can create clogging effects. In addition bacteria harmful to man can be injected into the systems. The primary consideration in the control of aquifer systems is to isolate the water from air exposure. EPA requirements (16) limit water treatment options. No inhibitors may be utilized; chlorine treatment is a possibility along with filters. The Texas A & M project used a sand filter for all water injection, and to date no problems are known to exist. The problems are similar to cooling tower water control. Injecting very hot water must be studied in detail. Is this the injection of a pollutant and what must be done to control the distribution? The potential savings of energy and capital which aquifer thermal energy can provide are expected to far outweigh the minor environmental effects.

Engineering Economics

The magnitude of the investment required to design and install a system including thermal energy storage in aquifers demands a thorough economic evaluation and justification. The large initial investment and extended benefit period associated with such systems make the analytical methods commonly known as "engineering economics" appropriate. Engineering economics provides the means to compare the relative economic merits of alternative systems in a variety of situations.

The economic justification of a system involving thermal energy storage in aquifers will generally include these four phases:

1. Estimation of system cost and benefits
2. Engineering economic analysis
3. Sensitivity analysis
4. Selection of the "best" alternative

Chapter XIII will develop these phases in detail and demonstrate their application with two case studies. Because all readers may not be familiar with the economic justification of major systems, the phases are introduced here so they may be recalled as the engineering details of the various systems are developed in subsequent chapters.

System costs are typically of two types--the initial capital investment which must be amortized or depreciated over the life of the system and the annual cost to operate and maintain the system. Accurate estimation of the initial capital investment is required to assure project design and construction within the budget approved by financial backers. Little cost data are available for the design, construction, and operation of systems involving thermal energy storage in aquifers; however, many contractors and engineers are familiar with the major components of the subsystems.

Subsystem	Major Components
Water Supply	Wells, pumps and storage tanks
Water Distribution	Piping and pumps
Energy Utilization	Water source heat pumps, solar collectors, cooling towers and chillers

Initial purchase cost and operating energy consumption data are available for each of these major subsystem components. System integration and control costs are more difficult to estimate and require experienced engineering judgment.

Energy saving is the major system benefit and most generally is estimated relative to the current system's energy usage or a proposed alternative system. Reasonable estimates of the energy savings potential can be made based on thermal loads and efficiency ratings. At present, the most difficult aspect of benefits estimation is the forecasting of future energy costs. Lacking any unforseen economic crisis or new technological breakthroughs, energy costs appear to be headed for increases substantially above the increase in total consumer prices. This point alone is a strong economic incentive for the consideration of systems involving thermal energy storage in aquifers.

The second economic justification phase uses the techniques of engineering economics to determine if the system's future benefits are sufficient to justify the required investment. Payback period is a simplistic approach used to evaluate a short-term investment problem. Payback period is inadequate when alternative systems have differing expected useful lifetimes or when the time value of money must be considered. Two general approaches which consider the time value of money or interest rate are demonstrated in Chapter XIII . First, the initial capital investment can be annualized using the capital recovery method to develop a uniform annual cost for the proposed system. Capital recovery can not readily handle special problems caused by income tax considerations and variable component lifetimes. The second method generates an annual cash flow for the system and determines the rate of return on investment (ROI) generated by that cash flow. ROI analysis is difficult without the aid of a computer or sophisticated calculator but provides the most comprehensive economic justification tool and can be applied to any cash flow pattern.

The results of an engineering economic analysis are subject to errors in the original estimates of system costs and benefits. Sensitivity analysis is a general technique used to determine the range over which those estimates can vary without changing the recommendation made from the analysis. Digital computer simulation is one means to test the sensistivity of a recommendation in an uncertain environment. An Energy System Investment Simulator (ESIS) program is outlined in Chapter XIII. The ESIS computes both the expected payback period and the ROI even when considerable estimation uncertainty is present.

The final economic justification phase is the selection of the "best" alternative. The word best is used with some caution because of the previously mentioned uncertainties. Usually, economic consideration alone is sufficient to make the final system selection decision. For example, the economic study may narrow a field of ten potential systems to two or three closely ranked "superior" alternatives. In other cases only two alternatives may have been considered and the results are in the form of future scenarios which favor each alternative, such as chosing system A if energy price increases are expected to be at least 15% per year, or otherwise chosing system B.

Thus, the final system decision must often be made using more than economic considerations. Corporate or community image might well be a key decision factor in the acceptance or rejection of novel energy systems. Other factors might include: risk, potential for future markets, energy availability, availability of investment funds, environmental considerations, and a host of other intangible factors.

References

1. <u>Proceedings of the Thermal Energy Storage in Aquifers Workshop</u>, NTIS No. LBL-8431-UC-94A, (1978).

2. Meyer, C.F., and Todd, D.K., "Conserving Energy with Heat Storage Wells," <u>Environmental Science and Technology</u>, Vol. 7, No. 6, (1973).

3. Meyer, C.F., Hausz, W., Ayres, B.L., and Ingram, H.M., Role of the Heat Storage Well in Future U.S. Energy Systems, NTIS, PB-263-480,(Dec. 1977).

4. Meyer, C.F., "Status Report on Heat Storage Wells," <u>Water Resources Bulletin</u>, Vol. 12, No. 2, (April, 1976).

5. Mathey, B., "Development and Resorption of a Thermal Disturbance in a Phreatic Aquifer with Natural Convection," <u>Journal of Hydrology</u>, 34 (1977), Elsevier Scientific Publishing Company, Amsterdam.

6. Werner, D., and Kley, W., "Problems of Heat Storage in Aquifers," <u>Journal of Hydrology</u>, 34 (1977). Elsevier Scientific Publishing Company, Amsterdam.

7. Jank, R., "Aquifer Storage in Germany," <u>Proceedings of the Thermal Energy Storage in Aquifers Workshop</u>, NTIS NO. LBL-8431-UC-94A, (1978).

8. Claesson, J., Eftring, B., Hellstrom, G., and Olanders, P., <u>Theortical Analysis and Computor Simulation of Solid-Fluid Heat Storage Systems in the Ground - Extraction of Heat</u>, Lund Institute of Technology, Sweden (1978).

9. Cormany, V. et. al., <u>Heat Storage in a Phreatic Aquifer</u>: Compuget Experiment, Electhecite de France (1978).

10. <u>Request for Proposals for Aquifer Thermal Energy Storage Program</u>, B-67756-A-G, Sept. 7, 1979, Battelle Pacific Northwest Laboratory.

11. Moltz, F.J., Warman, J.C., and Thomas, E.J., <u>Aquifer Storage of Heated Water: Experimental Study</u>, Auburn University (1978).

12. Ebeling, L., Reddell, D.C., Tostengard, S., Harris, W.B., and Davidson, R.R., "The Effect of System Size on the Practicality of Aquifer Storage," <u>Procedings of ISES Meeting</u>, Atlanta, Georgia (1979).

13. Todd, D.K., _Groundwater Hydrology_, Wiley New York, (1963).

14. Schaetzle, W.J., Brett, C.E., and Seppanen, M.S., _Heat Pump Centered Integrated Community Energy Systems_, NTIS No. ANL/CNSV-TM-25 (August,1979).

15. Schaetzle, W.J., Brett, C.E., and Ansari, J.M., "Thermal Energy Storage in Aquifers for a Solar Power Plant," in _Procedings of 2nd International Conference on Alternative Energy Sources_, Miami, (December 1979).

16. USEPA, "Interim Primary Drinking Water Regulations-Control of Organic Chemical Contaminents in Drinking Water," 43 Fed. Reg. 5756, (Feb. 9, 1978).

CHAPTER II

AQUIFERS AND GROUNDWATER

To understand the distribution of aquifers and the occurrence of groundwater it
is first necessary to have an understanding of the general structure of the Earth.
The parts of the Earth with which we have direct contact include three consecutive
layers termed the "lithosphere", the "hydrosphere," and the "atmosphere." The
lithosphere is the solid portion of the Earth and consists of the outermost rocks
with which we are all familiar and overlies the solid internal layers of the
earth. Surrounding the lithosphere is a layer of water known as the hydrosphere
which consists of the oceans, lakes, rivers, and groundwater. The outermost layer
of the earth is the gaseous envelope known as the atmosphere. All three of these
spheres interact both physically and chemically to create reservoirs for
groundwater and to influence the distribution of water to produce our principal
focus here, the aquifers (1).

Because the outer layers of the lithosphere are so strongly affected by the
chemical and physical activitites of both the hydrosphere and the lithosphere, the
rocks which ordinarily would be very dense are affected by decomposition and
erosion and are redistributed by movement of water. Consequently, the outer
lithosphere tends to be very porous and to contain cracks, crevices, and joints
which allow water to seep into and collect within the upper layers of the
lithosphere. Concentrations of this water are known as "groundwater" and the
rocks within which the concentrations are contained are known as "aquifers."

The classic concept of an aquifer, or that concept which is frequently preceived
by the uninformed, is that of an underground stream or river. In fact an aquifer
should be thought of as a saturated, rigid underground sponge. The aquifers are
generally rock layers which extend many hundreds of feet or many miles in all
directions. There are, of course, places on the Earth where underground rivers do
occur. These are generally in areas characterized by the presence of carbonate
rocks such as limestone which are highly subjected to solution, and contain
caverns such as the Mammoth Cave in Kentucky or the Carlsbad Caverns in New
Mexico. Groundwater of this type is not particularly effective for thermal
energy storage, but may be useful for water source heat pump applications.

The Hydrologic Cycle

Of all the fresh water that exists on the continents, 95 percent of it is
underground in the form of groundwater. Surface water, in the form of lakes,
streams and reservoirs is readily seen and appears to be a vast quantity of water,

yet it constitutes only five percent of the available fresh water resources. As a result of its being on the surface and readily accessible, it is often overused and misused. Surface water is also an integral part of groundwater reservoirs and in most places interfaces with groundwater and is indicative of the presence of groundwater (2).

Groundwater occurs naturally and comes principally from rain and snow which have been derived from the evaporation of surface waters, including the oceans, and which have precipitated on the Earth's surface (Fig. 2.1). Groundwater collects slowly over geologic time or in many cases may be renewed frequently and seasonally. During precipitation some of the water runs off and collects in ponds and streams, some of it evaporates, and where growing plants are present, some of it returns to the atmosphere by plant transpiration, and since the surface of the Earth is porous, some of it seeps down through the soil and becomes groundwater. Reducing runoff and enhancing infiltration of water into the Earth's surface is a subject of groundwater management. The continued process of evaporation and precipitation is called the "hydrologic cycle." Storage of water in aquifers is a temporary situation, geologically speaking, that is a normal part of the hydrologic cycle. As a part of the cycle, groundwater may move through the pore spaces of the lithosphere and eventually appear on the surface again at some lower elevation and become part of the surface water (3). These points of reentry are sometimes seen as springs or they may be points at which the water simply seeps into a stream or lake. This activity tends to maintain stream flow during dry periods. During extremely wet periods streams and lakes may fill up and the seepage may reverse so that the surface water seeps back into the aquifer, therefore becoming a part of the aquifer recharge cycle. This tendency for water to flow back and forth between the surface and the aquifers accounts for the aquifers maintaining their reservoir volume readily without a strong seasonal influence. The movement, occurrence, and distribution of water below the Earth's surface is known as "groundwater hydrology." Because water from precipitation enters the Earth at different temperatures depending on latitude, the average seasonal temperature of groundwater varies from north to south. These temperatures range from 72°F in the southern part of the United States to around 40°F in the northernmost part of the United States (Fig. 2.2). In general, aquifers in the mountain states tend to be cooler at the same latitude than aquifers elsewhere (4).

Where Groundwater Occurs

Pore Spaces

Groundwater collects in rocks that have cracks, crevices, and pore spaces. Because many rocks have been buried deeply in the Earth during their geologic history and are now at or near the surface, the pressure on them has been released, resulting in the presence of large numbers of joints and cracks. The quantity of groundwater stored in these rocks depends on the volume of pore spaces and on the number of joints and the extent to which they are open and connected. For the most part, jointed rocks do not provide much space for storage capacity, but water can move through them even if they are only slightly opened. (The degree of jointing varies from place to place and as long as highly jointed areas are interconnected with areas that are not quite so jointed, a useable reservoir may result). In limestone areas water moves along these joints and dissolves out relatively large spaces resulting in subterranean caves and even producing the popularly-conceived underground rivers. In many limestone areas of the world rivers can be seen to flow underground through surface openings in some places and flow out of the ground onto the surface in others (3).

Clastic rocks are rocks composed of the granular particles of preexisting rocks that have been broken up during geologic history, formed into sands, gravels, and clays, and then consolidated into rocks. In clastic rocks groundwater is stored

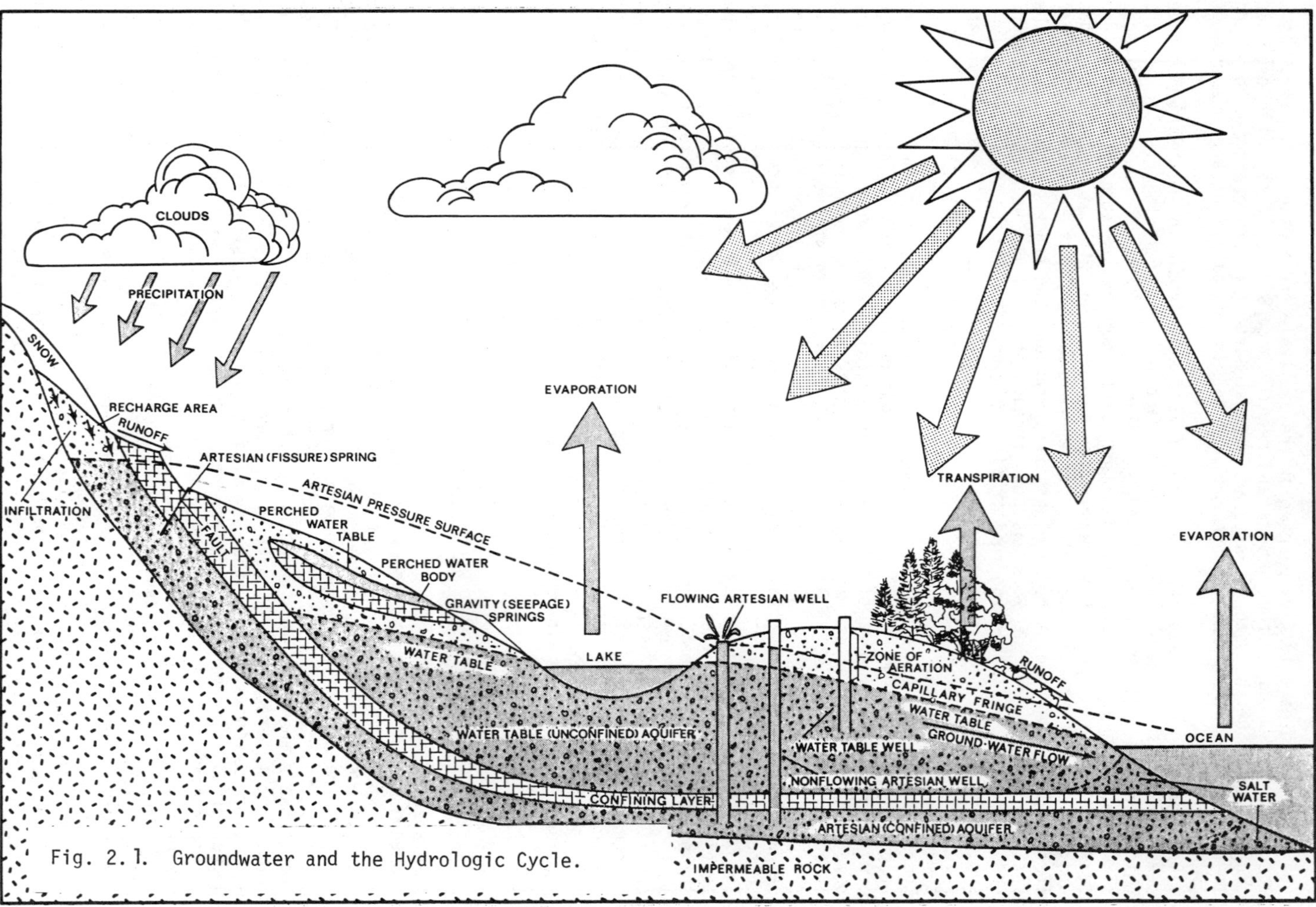

Fig. 2.1. Groundwater and the Hydrologic Cycle.

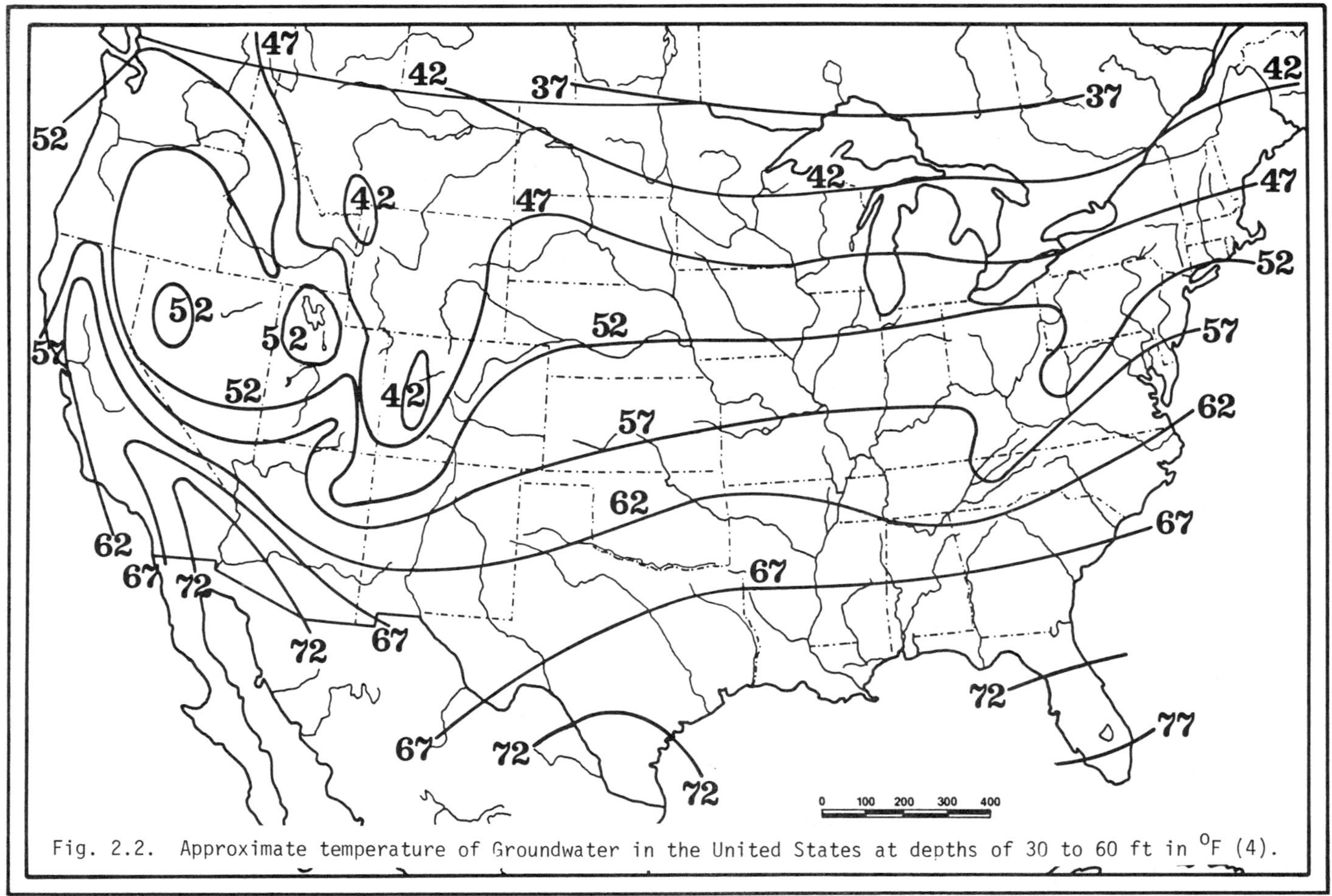

Fig. 2.2. Approximate temperature of Groundwater in the United States at depths of 30 to 60 ft in °F (4).

in the spaces between the grains. These pore spaces may constitute from five to 45 percent of the total volume of a rock. Pore space is at a maximum in rocks composed of well-sorted grains, meaning that the grains are nearly all about the same size. If a mixture of small and large grains is present the small grains will fit in the spaces between the large grains and partially fill the pores reducing the amount of space for groundwater.

Some of the grains in clastic rocks may be partially bonded together with cement such as calcium carbonate which further reduces the intergranular pore space and reduces the volume of groundwater which may be stored. So the amount of pore space available for storage is chiefly a function of the degree of sorting and the amount of cementing that has occurred during the history of the aquifer.

The amount of compaction also affects pore space in that a rock in which the grains fit tightly together has less pore space than one in which the grain arrangement is less compact. Compaction is usually a function of the age of the rock and the depth to which it has been buried in its history. Clastic rocks with their characteristic intergranular pore spaces are usually relatively chemically inert, being composed primarily of the mineral quartz, and ordinarily are described as sandstones or conglomerates. Shales are also clastic rocks but since they are composed primarily of clay materials the particles are so small that water adheres to them by capillary attraction and cannot move readily through the rock material. In some areas, particularly in regions such as Florida, clastic rocks may consist of limestone particles which have been created by physical and biological activities normally found in a coastal marine environment. Clastic limestones can be excellent groundwater reservoirs also if the pore spaces have not been adversely cemented by additional limestone.

Movement of Water

Porosity is the percentage of pore space in a rock and is a measure of the water-storage capability in a rock. Even though all the pores are connected in an aquifer and the aquifer is totally filled with water, not all of the water will move through the pore spaces, and, therefore, not all of it can be pumped out. Because water has a natural attraction to the surfaces of the grains in the aquifer, some of this water will always remain within the aquifer. This natural tendency to adhere to the surface of grains is called "capillary attraction." It is a very real force and has a large influence on the rate at which groundwater will move through an aquifer. Since wells and springs are supplied by water moving through pores we need to know something about the ease or difficulty with which the flow occurs. The ability of groundwater to flow through a porous rock is called "permeability," and as indicated above the percentage of pore space is "porosity." The ability of a reservoir to hold water is determined by the aquifer's porosity, but its ability to produce water is determined by its permeability. For the most part there is a more or less direct relationship between good porosity and good permeability (2).

If an aquifer has a porosity of 30 percent but will produce only half of this water, then the aquifer is said to have a specific yield of fifteen percent. A specific yield of fifteen percent, however, is a relatively large volume of available groundwater.

Subsurface Water and Groundwater

Water Table
The porous outer part of the lithosphere is termed "the zone of rock fracture." The pores or openings in this zone occur at different depths below the surface depending on the geologic setting. Water generally occurs within the porous areas but not all of this water is groundwater (5). At the surface is a "zone of

soil moisture" under which is a layer of openings partly filled with water clinging to the rock and soil particles. This is the "zone of aeration" (Fig. 2.1). At the base of the zone of aeration is the "capillary fringe" which consists of water drawn up into the subsoil from the water table by capillary action. Below the zone of aeration all the pore spaces are filled with water. The point at which all the pore spaces are full is the "water table" and below the water table is an area known as the "zone of saturation." The term groundwater refers properly only to water occurring in the zone of saturation. All the rest along with groundwater is called subsurface water and the rock formation in which saturation exists is called an aquifer. If an open hole or well is dug through the zone of aeration into the zone of saturation then the well will fill with water up to the level of the water table. The ground is usually moist in the zone of aeration, but it is improbable that water will seep out of the pores into the well from this zone. During rainy seasons the water table tends to rise and during periods of low precipitation it tends to fall. This variation can be observed easily in wells and results from recharge and discharge of the groundwater reservoir (2).

Discharge

Artificial discharge occurs when water is pumped from an aquifer through wells. Natural discharge occurs when groundwater emerges at the surface as a spring, or directly into a stream or lake and, finally, sometimes plants with long roots may reach into the aquifer and use some of the groundwater. Aquifer recharge results in the water level rising in wells. At some point associated streams will begin to enlarge, flow will resume in dry stream beds and flowing springs will develop, indicating that the aquifer has reached its capacity. Observation of the fluctuation of the water table level makes it possible to determine the amount of water that can be recovered from an aquifer on an annual basis. The discharge can then be limited to allow for net annual recharge and recovery.

Daming can be used to impound water associated with an aquifer and thereby insure long term recharge.

Effect of Well Pumping

Anytime a well is pumped the water table is drawn down around it to some degree in the shape of a cone. The depression formed in the surface of the water table around the well is called the "cone of depression." If there are other wells near a large well which is producing a large cone of depression, those wells will go dry when the cone of depression drops below the base of the well hole and other wells in the area may be lowered significantly reducing the amount of water that can be recovered (Fig. 2.3). A cone of depression can be in the order of magnitude of as much as a mile in diameter in large wells. The drawdown of a large well and its adverse effect on neighboring wells is frequently the basis for law suits. In the thermal energy storage in aquifers technique, the cone of depression is limited significantly because the groundwater is recycled through a second well, thereby continuously recharging the aquifer with a net consumption of zero. Systems for energy storage should be planned so that wells drawing water for consumptive purposes are not associated with the storage volume of the aquifer.

Perhaps the most difficult situation occurs when a region becomes dependent on a large number of wells and an aquifer eventually is pumped until artifical discharge exceeds recharge. In this case a detailed study must be made to determine if well pumping must be controlled or restricted or if it is more economical to limit the drilling of new wells and to replace the old wells with different systems such as deeper wells.

Again, the use of aquifers for thermal storage has no net effect on the recharge/ discharge picture in a particular region, but it is extremely sensitive to the comsumptive uses of groundwater in the aquifer in the surrounding area.

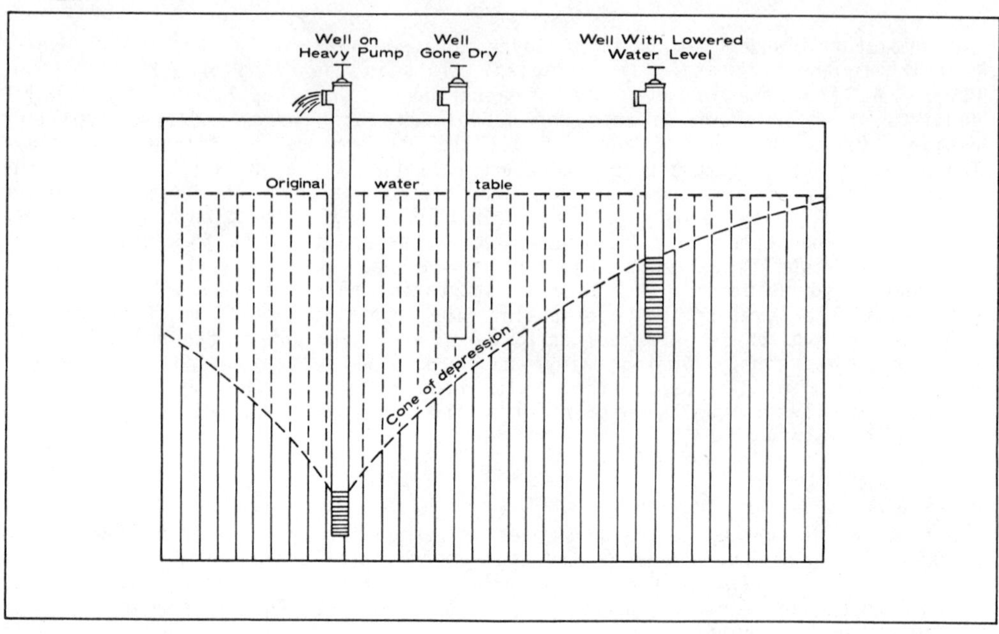

Fig. 2.3. Cone of Depression.

Types of Aquifers

Unconfined Aquifers

To this point we have been talking about a situation in which water infiltrates down through the Earth's surface through a more or less homogeneous layer of porous rock material until it reaches a non-porous layer and stops migrating. This kind of aquifer, found mostly at shallow depths within the Earth's surface, is termed an "unconfined aquifer" (Fig. 2.1). Unconfined aquifers are found in most areas where there is abundant rainfall. The capacity that can be expected from an unconfined aquifer varies considerably depending on precipitation in the area and the geological setting of the aquifer itself. Sometimes unconfined aquifers are not homogeneous, as in the case of flood plains adjacent to river systems. For instance groundwater frequently collects in ancient river channels which have long since been buried and which consist largely of sands and gravels. As a result the aquifer itself may tend to occur in a serpentine pattern similiar to the normal pattern of the stream associated with it making it possible to drill into the aquifer at one point but go only a few tens of feet away and miss it with a drill entirely. This does not mean that there are not large quantities of water available in the area. Use of groundwater for thermal energy storage in this setting requires some exploration to define the geometry of the aquifers.

It also occurs in some places that layers of clay within the aereated zone may create the formation of a "perched" water table. In such cases the capacity generally is very low and a well can be pumped dry quickly. Subsurface water infiltrating down to a perched water table will flow laterally until the edges of the clay layer are reached and then continue to flow downward until the permanent water table is reached. If a perched water table is intercepted the permanent water table can usually be reached by continuing to drill deeper.

In an unconfined aquifer, when the upper limit of the aquifer is the top of the zone of saturation, groundwater is said to be under water table conditions. The aquifer is called a water table aquifer. In an unconfined or water table aquifer the pressure at any level within the aquifer is equal to the depth from the water table and may be expressed as hydraulic head in feet of water.

Confined Aquifers

Pervious and Impervious Beds

In some areas where sedimentary rocks are predominant they exist in layers, each layer of which may be tens to hundreds of feet in thickness. Some of the layers may lack porosity and be essentially impervious to water movement and are, therefore, confining layers, and some layers are sufficiently porous to collect and transmit groundwater through them, and are, therefore, aquifers. When the edge of a layered aquifer is exposed at the surface, it then becomes a recharge area and is the source of infiltrated water for the aquifer. A confining layer may occur above and below the aquifer so that the water may be transmitted through the aquifer into deeper areas of the Earth's surface creating an aquifer which is under hydrostatic pressure. When an aquifer such as this is sandwiched between impervious beds, the aquifer and the groundwater in it are said to be "confined" (Fig. 2.1).

Artesian Pressure

When wells are drilled through a confining layer into a confined aquifer, the water in the well will rise to a point somewhere above the top of the aquifer. The well then is known as an "artesian well." The elevation of the water in the well at this point is known as the piezometric level of the aquifer. By contouring the elevations of a series of artesian wells in an area, a piezometric surface can be described which is similiar in concept to the water surface of a water table aquifer. The piezometric surface is sometimes also called the "artesian pressure surface" (Fig. 2.1). In an artesian aquifer the pressure within the aquifer is equal to the depth from the piezometric level and is expressed as hydraulic head in feet of water. Sometimes the artesian pressure surface may actually occur above ground level; thus, when wells are drilled into the artesian aquifer, water will flow freely out of the well under it own pressure creating a flowing artesian well. If the artesian pressure surface is below the ground surface, water will rise to the level of the pressure surface and the well is still described as an artesian well, but, of course, is not a flowing well.

If the piezometric surface is well defined in a particular area it then becomes possible to predict to what level water will rise in any new well that is drilled in the area. Aquifer management for confined aquifers is similiar to management for water table aquifers. There is a natural balance between recharge and discharge in confined water systems, but if the system is discharged artifically in excess of the aquifer's ability to recharge, then the piezometric surface will fall and if it falls below the land surface, flowing artesian wells will cease to flow. This means, then, that pumps must be added to the wells and the user then must begin paying for energy when using the water.

Commonly, aquifers consist of rock of sedimentary origin, such as limestones, dolomites, siltstones, sandstones, and conglomerates although, in some instances, fractured and jointed metamorphic and igneous rocks may contain water in sufficient quantities to serve as a source of fresh water. Aquifers adequate for thermal energy storage underlie approximately 60 percent of the land area of the continental United States (Fig. 2.4). Of major importance are the areas adjacent to major river basins which generally afford suitable aquifers associated with the rivers. The Mississippi, Missouri, Ohio, and Allegheny systems are examples of such basins. Major population centers are located along these stream basins. Our

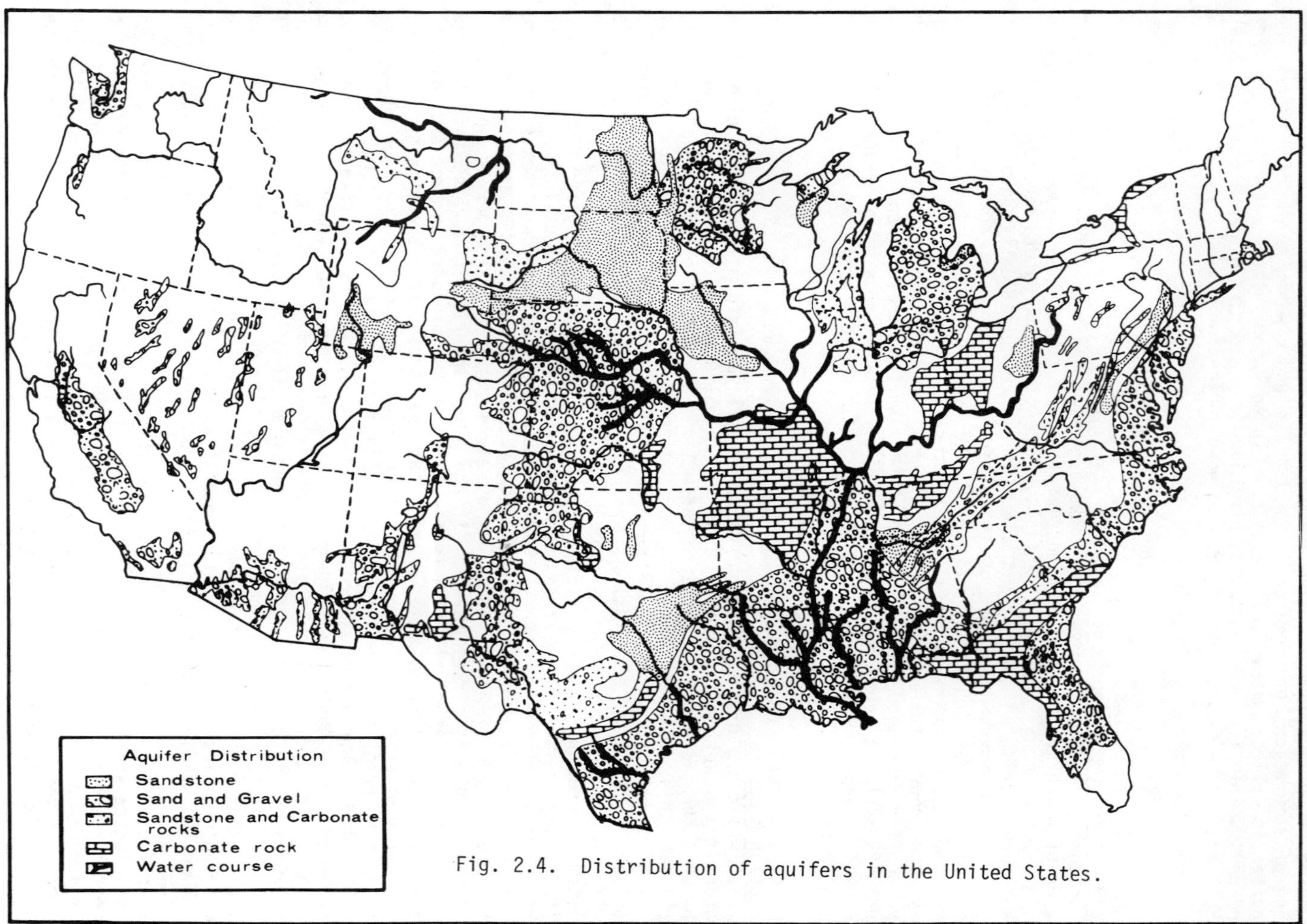

Fig. 2.4. Distribution of aquifers in the United States.

present population centers were located historically because of the need for aquatic transportation, especially along the rivers, lakes, and the sea coast. As a result over 75 percent of the population has access to adequate aquifers. Comparing the distribution of population to distribution of aquifers, it is estimated that use of thermal energy storage systems is feasible in 75 percent of the metropolitan areas and 60 percent of the non-metropolitan areas.

Carbonate Aquifers

Carbonate aquifers are unique. Carbonate rocks found near the surface are usually jointed as are other rock types. Water infiltrating carbonates normally contains carbon dioxide in the form of carbonic acid. Carbonic acid along with organic acids from overlying soils dissolve the surfaces of the cracks and crevices through which they pass and creates irregular passageways. Where large concentrations of joints, cracks and crevices occur along the fractured zones, solution will be most profound. In those areas, after a time cavernous openings are dissolved out of the rock creating the familiar features such as the underground caverns mentioned earlier. In such areas sometimes rivers flow into the underground and rivers elsewhere emerge from it. Caves can be found in almost any terrane which is predominantly limestone (6). Thermal energy storage in limestone areas has little likelihood of success because of the relatively rapid migration of water through the limestone systems.

Aquifers in Clastic Rocks

Sandstone Aquifers

One of the prominent clastic artesian aquifers in North American is the Dakota Sandstone which has its recharge area along the east front of the Black Hills (Fig. 2.5). The sandstone flattens out under the Central Plains and is overlain by impervious shales that create a conduit for many miles to the east. The Dakota Sandstone is recharged by streams flowing from the Black Hills across its surface outcrop. Early in the history of drilling water wells in this area, artesian wells frequently spouted water over 100 feet in the air in the eastern parts of the area. Over a long period of time many of these flowing wells were allowed to run continuously forming small lakes in low areas. After a few years the artesian pressures dropped and eventually many of the wells ceased to flow. Groundwater management techniques were applied by the U.S. Geological Survey to attempt to recharge the Dakota Sandstone, but reservoir pressures have not increased significantly so far. Large aquifer systems of this type are available in most areas of the Great Plains region of the United States and, when intercepted by drilling, will usually produce signficant quantities of water and even artesian flow. However, these aquifers are fairly deep in most places and the expense of drilling to them must be offset by the need for the water.

Coastal Plains

The Coastal Plains of the Atlantic and Gulf regions are two of the most productive groundwater provinces in the United States (Fig. 2.6). The Coastal Plains extend several hundred miles inland from the present shoreline and are areas where the Atlantic and Gulf seas have overlapped the continent during past geologic time and left extensive layers of clastic deposits such as sand, silt, clay, and some limestones. In these areas only the limestones are highly consolidated so that the limestones provide confining beds for the clastic deposits and create confined aquifers which in some areas are artesian. Broad expanses of clays also provide confining beds and in some areas the limestones may be permeable and also produce water. Salty water is produced occasionally and is

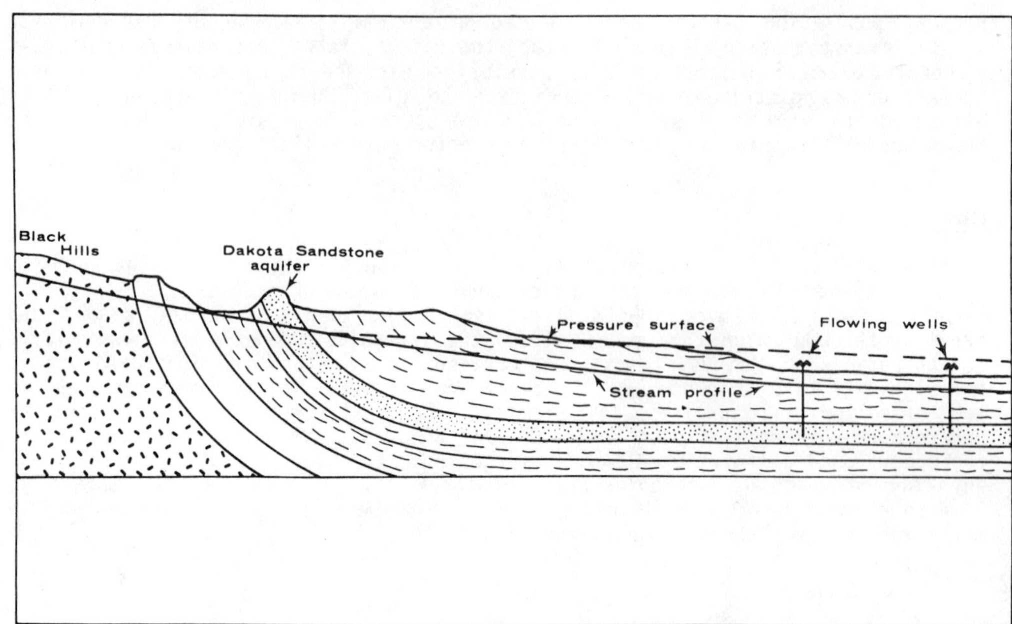

Fig. 2.5. A sandstone aquifer with flowing wells.

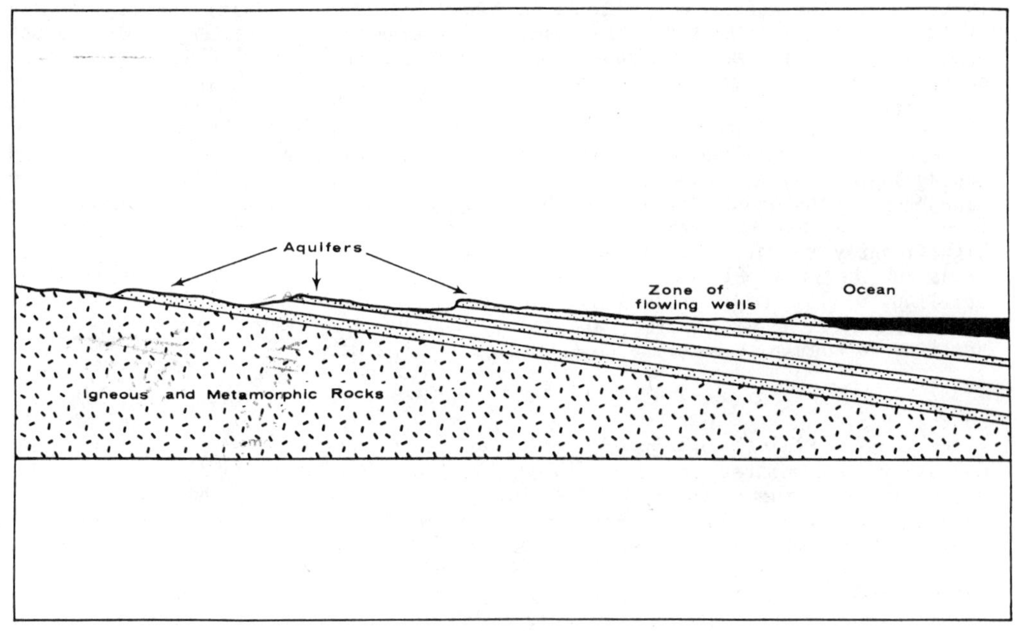

Fig. 2.6. Coastal Plain Aquifers.

probably seawater deposited in the sediments at the time of their formation which has been stored there ever since. In specific areas such as southeastern Florida some of the most permeable aquifers occur. These occur in limestones.

Glacial Deposits

One of the more favorable areas for the presence of shallow aquifers in the United States is in the area extending from Nebraska eastward across the northern part of the United States. These areas in past history have been covered by continental glaciers. Broad expanses of clastic materials were left there by glacial activity. Glacial deposits consist of unconsolidated sands, gravel, boulder clay, and clay deposits in a varied arrangement of forms.
Some of the material, most of which is very porous, was deposited along the lakeshores associated with the great ice sheets, some are river channel sands, and others are broad sheets of clastic materials that were deposited by meltwater far beyond the fronts of the continental glaciers. Most of the wells in these areas must be pumped. The deposits themselves are several tens of feet thick and lie upon older rocks which are essentially impermeable. This creates a condition forming an ideal water table well and produces vast reservoirs of groundwater over broad areas. In some places in the glacial areas clay may predominate and wells produce very little water. Aquifers in glacial areas are characterized by excellent flows without much draw-down of the pressure surface. The use of glacial deposits for aquifer thermal storage is very promising.

Alluvial Fans

In many mountain areas of the United States and particularly in the arid regions, the mountains themselves are characterized by having an apron of eroded clastic materials surrounding the base of the mountain. These structures are known as "alluvial fans" (Fig. 2.7). Alluvial fans are found in New Mexico, Utah, Arizona, Nevada, and Eastern California. They are characterized by having layers radiating away from the mountains which are wedge shaped being thinner near the mountains and thickening in all directions away from the mountains. The aquifers in the alluvial fans are relatively deep in the upper portions of the fan. In the lower parts of the fan, however, the same aquifer, of which there may be many in the fan structure itself, are closer to the surface and accessible. In the intermediate parts of the fan slopes, the aquifers tend to be artesian and in some cases may even flow. In the valley areas below the alluvial fans the sediments become very fine and lack sufficient permeability to allow water to pass through them. Aquifers in alluvial fans are usually drilled for culinary water and are heavily used. Therefore, the use of alluvial fan aquifers may be somewhat limited for aquifer thermal storage because of conflicts in aquifer management.

Flood Plains

Rivers are frequently surrounded on either side and underneath by clastic deposits derived from the normal river processes such as seasonal flooding and meandering. Sediments are transported from the upper reaches of the stream where erosion dominates to the lower flood plain areas where deposition of its sedimentary load dominates. Rivers tend to shift seasonally depending on rainfall from dominantly erosive to dominantly depositional almost anywhere along their course. As a result the more level and lower reaches of a stream tends to be flanked on both sides by relatively thick sections of mixed clastic deposits composed of gravel, sand and clay (Fig. 2.8). Geologically these deposits are relatively young and have not had time to become hardened into rock. They almost universally contain groundwater and usually make excellent aquifers because of the predominance of coarse clastics and the stabalizing influence of the associated river.

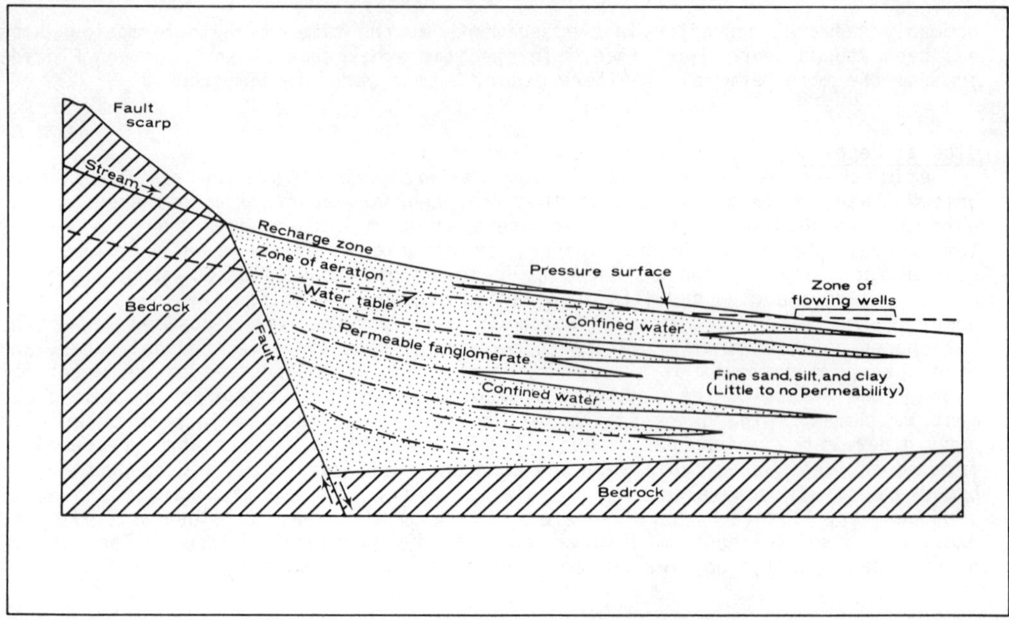

Fig. 2.7. Alluvial fan aquifers.

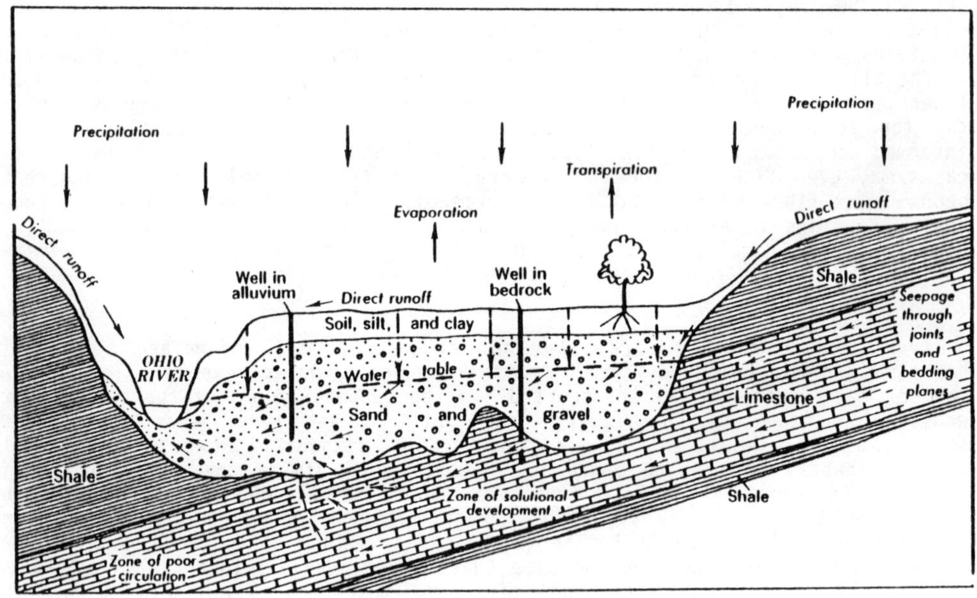

Fig. 2.8. A flood plain aquifer.

However, because the coarse materials in flood plain deposits are deposited in specific riverine environments, such as bends (meanders) of the river, the aquifers produced by riverine processes can have complex geometries and be difficult to intercept with a drill. In the course of its geologic history a river tends to wander around with its meanders (bends) moving both downstream and laterally leaving coarse sedimentary deposits that are discontinuous, tending to be serpentine or crescent shaped mirroring the shape of the river, and being separated by clay deposits that act as confining barriers. These deposits can be relatively thick in places (10's to 100's of feet) with the serpentine aquifers stacked randomly in a complex three-dimensional geometry throughout the breadth of the flood plain. Some of the aquifers may be interconnected, some not. Where the aquifers are interconnected, vast reservoirs may be formed with very large volumes of available groundwater. These are the kinds of aquifers that occur along the major rivers of North America and where many of the larger cities are located (Fig. 2.8)

Flood-plain sediments are deposited on pre-existing surfaces which may have considerable topographic relief. Consequently, it may be desireable to conduct a seismographic survey prior to drilling to locate the thickest sections of these "modified unconfined aquifers," and to avoid drilling in thin sections where the underlying bedrock relief is high and the probability of a good aquifer is low.

Groundwater Law

Although water use has been regulated in the United States for many decades the new and innovative uses for groundwater are creating today a whole new field in water law and policy. The biggest problem is the multifunctional use of groundwater for both consumptive and nonconsumptive purposes. In the past, most of the problems in groundwater utilization resulted from increases in consumption through time associated with growing populations and industrialization. Consequently, there has been more demand than available supply and increasing litigations have resulted.

Basic problems lie where water is used for competing functions which are not compatible and where competition for use of the water for the same function needs to be resolved (7). This is complicated by a jurisdictional problem resulting from the Constitution of the United States, and relates to the division of power between the federal and state governments. The Reclamation Act of 1902 and the Desert Land Act of 1877 are the legislative references that have proven most important in water-resources management in the United States. In dealing with groundwater the courts tend to define groundwater in two categories: underground streams and percolating waters. For the most part the courts seem to consider all subterranean water as percolating water unless it can be shown that it moves in a defined channel. Percolating waters are covered under three different concepts of rights: (1) the common law doctrine of riparian rights (2) the appropriative doctrine and (3) the reasonable-use rule.

The riparian rights principal is based on the concept that the owner of the land has the absolute right to use all minerals found on the land including groundwater. The appropriative doctrine is one in which the rights are acquired on a time-dependent priority basis regardless of who owns the land. In the reasonable-use rule the land owner must consider equally the rights of all other owners of land overlying the same groundwater supply and must make only reasonable use of the percolating waters underlying his land. In the doctrine of riparian rights, the right is retained to the use of the groundwater but the right to the groundwater is not lost due to non-use. Under the doctrine of appropriation an individual cannot acquire rights to groundwater for nonbeneficial use and in turn, his rights, once established, are lost if use of the water is discontinued. Today seventeen western states recognize this doctrine to some extent. In 1877 the Desert Land Act recognized the local doctrine of appropriation, leaving it up to each state to determine for itself to what extent the rule of appropriation or riparian rights should apply. The U.S. Reclamation Act of 1902 reconfirmed the Desert Land Act by stating "vested rights to the use of water from a stream or public land for mining, shall be maintained and protected."

In the future further legal complications will arise by the creation of a resource resulting from thermal energy storage in aquifers. When groundwater is either heated or chilled a resource is created which has a much higher economic value. Litigation could result if a conflicting use of the aquifer in an adjacent area caused a change in groundwater temperature. These kinds of legal problems must be anticipated in each State to prevent unnecessary legal entanglements where groundwater thermal storage systems are used.

References

1. Longwell, C.R., and Flint, R.F., Introduction to Physical Geology, John Wiley and Sons, Inc., New York, New York, (1962).

2. Water Well Drillers Beginning Training Manual, National Water Well Association, Worthington, Ohio, (1971).

3. Eardley, A.J., General College Geology, Harper & Row, New York, New York, (1965).

4. Collins, W.D., Temperature of Water Available for Industrial Use in the United States, U.S. Geological Survey Water-Supply Paper 250-F, Washington, D.C., (1925).

5. Todd, D.K., Ground Water Hydrology, John Wiley & Sons, Inc., New York, New York, (1959).

6. Shelton, J.S., Geology Illustrated, W.H. Freeman and Company, San Francisco, California, (1966).

7. DeWeist, R.J.N., Geohydrology, John Wiley & Sons, Inc., New York, New York, (1965).

CHAPTER III

AQUIFER SYSTEM ANALYSIS

Introduction

The analyses of aquifer systems that follow are limited to considerations of
one-phase flow, water only, in saturated zones (1,2). The voids in a saturated
aquifer are completely filled with water. However, the overlying unsaturated
zones may contain soil moisture which can interact with the saturated zone. The
interaction between the two zones is of no concern as most energy storage projects
will utilize sets of injection and withdrawal wells completed in a saturated
aquifer with the water and thermal energy being transferred between the sets of
wells.

In general, physical properties of individual aquifers may vary widely. For this
chapter, properties of the individual energy storage system and fluid flow system
are assumed to be constant. Aquifers analyzed are assumed to have constant
thickness and to be homogeneous. These assumptions are a fair approximation of
real conditions in most cases. Only a relatively small portion of a large aquifer
is normally utilized for energy storage. Engineering corrections can be
approximated for variations in thickness and homogeneity. Aquifers may range in
thickness up to a few hundred feet and extend for a number of miles.

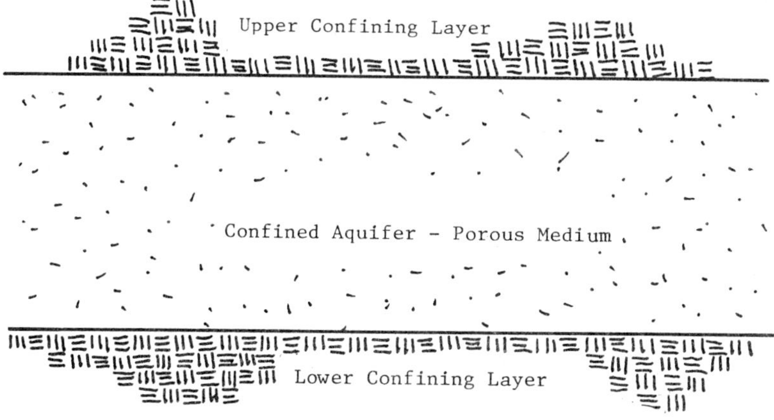

Fig. 3.1. Schematic of confined aquifer.

Numerous aquifers exist near the Earth's surface. Aquifers occurring mostly within a few hundred feet of the surface, but some at depths of a thousand or more feet, are considered for thermal energy storage. In general, aquifers will have a uniform thickness and sizable lateral extent. A layer of non-permeable material confines some aquifers both above and below. This is a confined aquifer as discussed in Chapter II. The thickness of the aquifer is the vertical distance between the confining layers (Fig. 3.1).

Aquifer Fluid Flow Properties

Porosity (ϕ) is defined as the volume of voids or pore space within a rock body with respect to the bulk volume of the rock and is expressed in terms of a fraction or as percent. Part of the voids in a rock body are isolated, not interconnected with the voids system, and hence are not available for the storage or transmission of fluids. The volume of the interconnected pore space is referred to as effective porosity. Inasmuch as effective porosity is that measured in the laboratory and is operative in physical responses in nature, effective porosity of an aquifer is called simply "porosity". The porosity is:

$$\phi \equiv \frac{V_v}{V_o} = \frac{V_w}{V_o} = v_w \tag{3-1}$$

where

$$
\begin{aligned}
V_v &= \text{void volume} \\
V_o &= \text{total volume} \\
V_w &= \text{water volume, and} \\
v_w &= \text{volume of water per unit volume}
\end{aligned}
$$

The solid fraction "f_s" is:

$$f_s = 1 - \phi \tag{3-2}$$

In the zone of saturation of an aquifer, the pore space is filled with water, so that the percent porosity represents the fraction of the bulk aquifer containing water, or the volume of contained water.

Porosity of rocks ranges from near zero to some 45 percent. Porosity of water-bearing rocks is likely to be from about five percent to 30 percent, commonly ten to 30 percent. Porosity in the range of 20 percent to 25 percent is considered "good". Methods of measuring porosity are discussed in Pirson (6) and Scheidegger (7). Table 3.1 gives the porosity range for a number of substances.

TABLE 3.1 Porosity of Various Aquifer Materials

Formation	Location	Porosity - Percent		
		Aver.	Min.	Max.
Sandstones	Wyoming	19.7	8.8	27.0
Limestone	Wisconsin	2.9	0.5	6.7
Fine Sand	California	46.2		
Mud	Hudson River	88.2		
Very Fine Sand		47.7		
Sand-Silt-Clay		88.2		
Bartlesville Sand	Oaklohma	18.3	7.6	32.0

Compiled from data in Reference 8.

The degree of saturation is defined as the volume of water to the void volume in a porous rock or:

$$S \equiv \frac{V_w}{V_v} \tag{3-3}$$

Again, for the included analyses, the water volume equals the void volume, that is, the aquifer is considered to be saturated.

The specific weight of an aquifer unit volume is:

$$\gamma_a = \frac{W_t}{V_o} = \gamma_{water} \, \phi + \gamma_{rock} \, (1 - \phi) \tag{3-4}$$

or the density is:

$$\rho_a = \rho_{water} \, \phi + \rho_{rock} \, (1 - \phi) \tag{3-5}$$

<u>Water Flow Rate and Velocity</u>

The velocity in an aquifer is hard to define because actual velocities vary with porosity and location. As a result the flow in volume per unit time is normally utilized. This can be the total flow from the well penetrating the aquifer or the flow per unit thickness of aquifer. Many calculations and curves are normalized per unit thickness of aquifer.

The flow into or from a well can be given as:

$$Q \text{ in } \quad ft^3/day, \; ft^3/hr, \; ft^3/min, \text{ or}$$
$$gal/day, \; gal/hr, \; gpm, \text{ or}$$
$$m^3/day, \; m^3/hr, \; m^3/min, \; 1/sec \text{ or other unit}$$

Most common usage is gpm or m^3/hr which are similar in magnitude.

$$1 \text{ gpm} \quad = \quad 0.227 \; m^3/hr \quad = \quad 13.6 \; m^3/day$$

The normalized flow rate per unit thickness of aquifer is:

$$Q/b = 1 \text{ gpm/ft} \quad = \quad 0.745 \; m^3/hrm$$

The velocity in the aquifer must be an average number as no cross section is ever homogeneous. As a result velocity is rarely used in geological evaluations. A velocity defined by dividing the flow rate Q by aquifer cross-sectional area A is known as the specific velocity V_S or:

$$V_s = Q/A \tag{3-6}$$

A more appropriate expression of velocity or "true velocity" assumes the porosity of a cross-sectional area is identical to the calculated porosity of the aquifer. This results in:

$$V = V_s/\phi = Q/A\phi \tag{3-7}$$

The true velocity represents the average water front velocity of flow in an aquifer or the average velocity of the volume flow. In a preponderance of analyses the volume flow rate is utilized.

Water Flow

Water flow is a function of pressure distribution and physical properties of the aquifer. These quantities can vary widely even in short distances. For the purposes of this study these properties are assumed to be constant in the area of analysis. In general, flow is proportional to pressure gradient (head gradient) and area (1,2,3,4,5) or:

$$Q \propto A \frac{dh}{dl} \qquad (3-8)$$

This flow relationship is known as Darcy's Law. Darcy first derived the relationship in 1856 (3). The equation can be developed from basic fluid mechanics. Most flow in an aquifer, except very near a well, is in the range of a few feet per day or a very small fraction of a foot per second. Also, the pore spaces are very small, consequently the Reynolds Number is also small. For this range of Reynolds Number, flow is in the region where pressure losses are proportional to velocity.

The constant of proportionality is called permeability (K) or:

$$Q = KA \frac{dh}{dl} \qquad (3-9)$$

The units of K are velocity. The common English units for the above equation are Q in gpd, K in gpd/ft^2, A in ft^2 and dh/dL in ft per ft. Many systems of units are used in the above equation. K is also known as hydraulic conductivity, effective permeability, coefficient of permeability, and seepage coefficient.

The most common unit for K in the United States for water movement is the flow of 60°F water in gallons per day through a cross section of one square ft for a hydraulic gradient of one ft per ft. These are the units normally used for tests in the laboratory. The U.S. Geological Survey has named this unit the "meinzer" in honor of the late Dr. O.E. Meinzer.

A more general form of Darcy's Law is:

$$Q = \frac{k\gamma}{\mu} A \frac{dh}{dl} \qquad (3-10)$$

where k is the specific permeability and normally given in darcys. Solving the equation for specific permeability gives:

$$k = \frac{\mu Q}{\gamma A (dh/dl)} = \frac{Q}{A} \frac{\mu}{dp/dl} \qquad (3-11)$$

where μ is fluid viscosity, and γ is specific weight.

The normal units for the darcy are in cgs units or:

$$1 \text{ darcy} = \frac{\text{centipoise } cm^3/\text{sec } cm^2}{\text{atmosphere}/cm} \qquad (3-12)$$

The relationship between specific permeability and the laboratory coefficient of permeability for water at 60°F is:

$$1 \text{ darcy } (k) = 18.2 \text{ gpd/ft}^2 \ (K) \qquad (3-13)$$

There is no direct correlation between permeability and porosity. For example, large sand grains with relatively large passages would have the same porosity as small sand grains if the grain shape and packing are identical. Grain size is frequently directly related to permeability as the permeability of gravel in most cases is much higher than that of sand.

Relative values of permeability for various types of aquifers are given in Table 3.2. Gravels have the highest permeability with permeability decreasing as the sands become finer. Aquifers with a permeability of 10 gpd/ft^2 or higher are generally well suited for thermal energy storage. The use of an aquifer with values of permeability from 10 to 10^{-3} gpd/ft^2 is marginal because pressure losses through the aquifer become excessive. Permeability in a single aquifer can vary appreciably. Several orders of magnitude of change (factor of 100) is not uncommon. For the purpose of calculation, aquifers in general are assumed to be homogeneous with constant values of porosity, thickness, and permeability. Engineering evaluations must be made on overall systems. Even when good geological data are available, test data are normally required for aquifer sections in each well.

TABLE 3.2 Approximate Aquifer Permeability Ranges

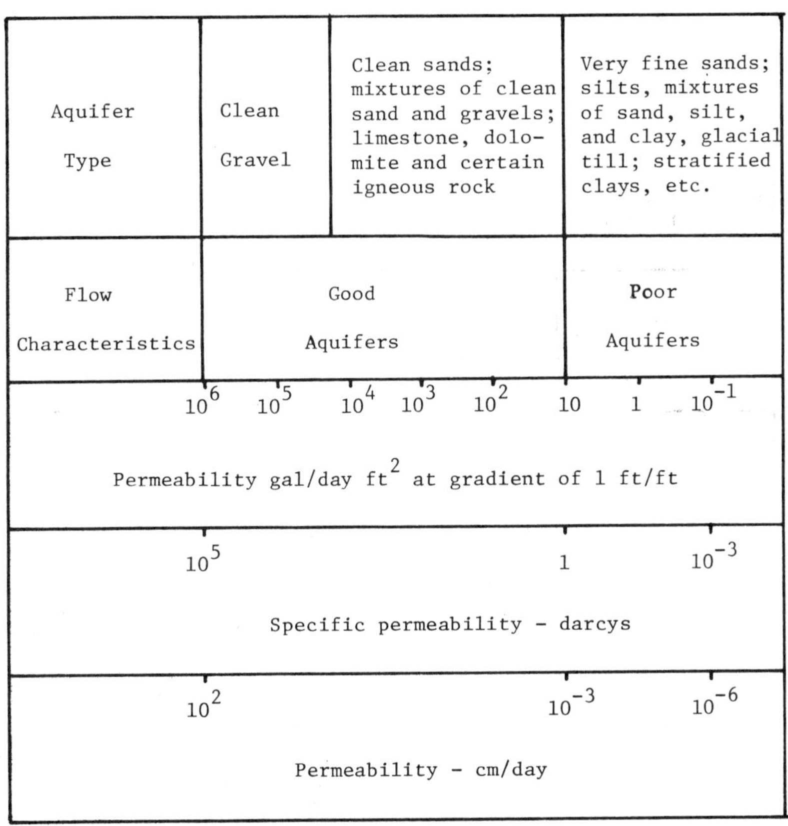

Aquifer Type	Clean Gravel	Clean sands; mixtures of clean sand and gravels; limestone, dolomite and certain igneous rock	Very fine sands; silts, mixtures of sand, silt, and clay, glacial till; stratified clays, etc.
Flow Characteristics	Good Aquifers		Poor Aquifers

Natural Flow in Aquifers

Natural flow in an aquifer is subject to the equations of flow. The natural flow in an aquifer can be stated as:

$$Q/A = K \text{ grad } h \tag{3-14}$$

For most natural flow, the gradient of h is gradual and the flow rate can be estimated. If large solution channels or major geological discontinuities exist, the calcualtions become questionable.

Figure 3.2 shows a schematic cross-section of an aquifer indicating a variation in groundwater level or equilibrium head known as the "piezometric surface." The piezometric surface represents the pressure level at any point in the aquifer. The water level in non-flowing wells rises to the piezometric surface. The reference head in the aquifer at any point is the height above a horizontal reference plane of the piezometric surface at that point. The location of the reference plane is of no consequence as the flow is a function only of the gradient of head (pressure).

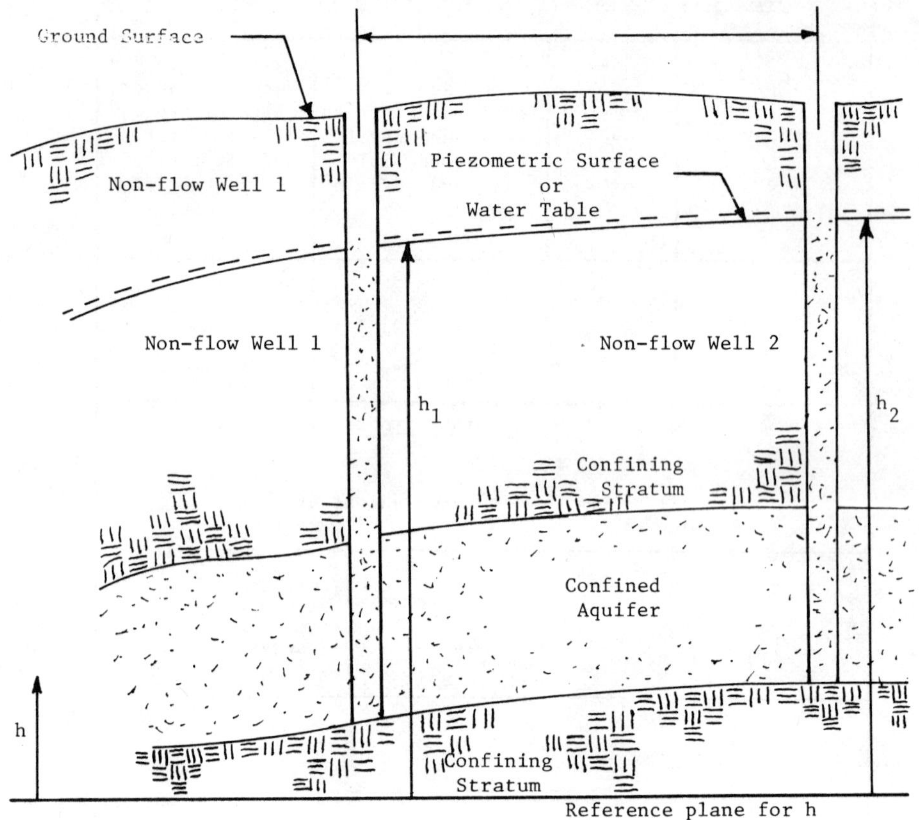

Fig. 3.2. Schematic of general aquifer with piezometric surface.

If the wells (relatively flat country) in Fig. 3.2 are 5,000 ft apart, the difference between h_2 and h is two ft, and the permeability is 10 gpd/ft^2, the natural flow from well two to well one is:

$$Q/A = 10\text{gpm/ft}^2 \cdot 2\text{ ft/5000 ft}$$
$$= 0.004 \text{ gpd/ft}^2$$
$$= 1.46 \text{ gal/yr/ft}^2$$

The specific velocity as defined by equation 3-6 is:

$$V_s = Q/A = 0.004 \text{ gpd/ft}^2 \; 1\text{ft}^3/7.48 \text{ gal}$$
$$= 0.000534 \text{ ft/day}$$
$$= 0.195 \text{ ft/year}$$

The velocity as defined by equation 3-7 for a porosity 0.3 is:

$$V = V_s/\phi = 0.000534 \text{ ft/day}/03$$
$$= 0.00178 \text{ ft/day}$$
$$= 0.65 \text{ ft/year}$$

A permeability of 100 gpd/ft^2 will increase the numbers by a factor of ten.

Natural flow rates normally range between five ft/yr to five ft/wk. To consider an aquifer for thermal energy storage the flow rates should be fifteen ft/yr or less. Aquifers with larger flow rates can be utilized but caution must be exercised for estimating natural flow.

Wells

Wells can act either as a sink (water withdrawal) or a source (water injection) to an aquifer. The ideal flow in a homogeneous aquifer can be approximated as potential flow. This can be shown mathematically (8,9) and empirically from data collected from actual well flow patterns (10,11).

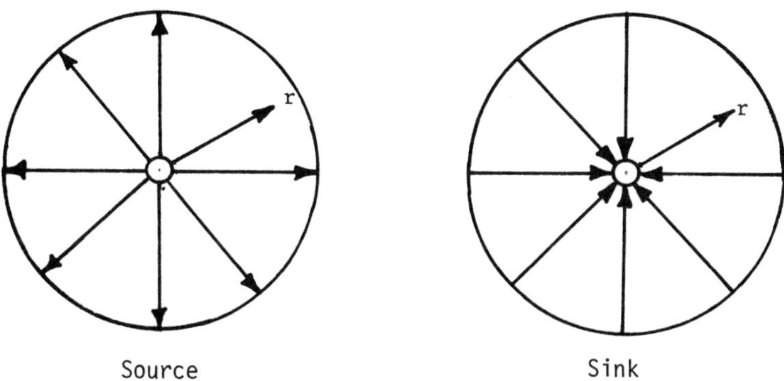

Fig. 3.3.   Independent source and sink flow.

The simplest forms are the independent sink and source flow. The streamlines for these systems are shown in Fig. 3.3. The flow is directly from the source (well) or directly into the sink (well). The velocity at any radius with 100 percent porosity from continuity considerations is:

$$V = \frac{q}{2\pi r} \quad (\text{Source}) \tag{3-15}$$

$$V = -\frac{q}{2\pi r} \quad (\text{Sink}) \tag{3-16}$$

where q is the flow into or from the aquifer per unit of aquifer thickness and r is the distance from the center of the well. If the porosity is less than 100 percent, the velocity is:

$$V = \frac{q/\phi}{2\pi r} \quad (\text{Source}) \tag{3-17}$$

$$V = -\frac{q/\phi}{2\pi r} \quad (\text{Sink}) \tag{3-18}$$

The term q/ϕ or corrected flow is used, since only a fraction of the volume is available for water transport. The value is also used for several later calculations. This also shows that the velocity adjacent to the well is very high and the velocity decreases as the distance from the well increases.

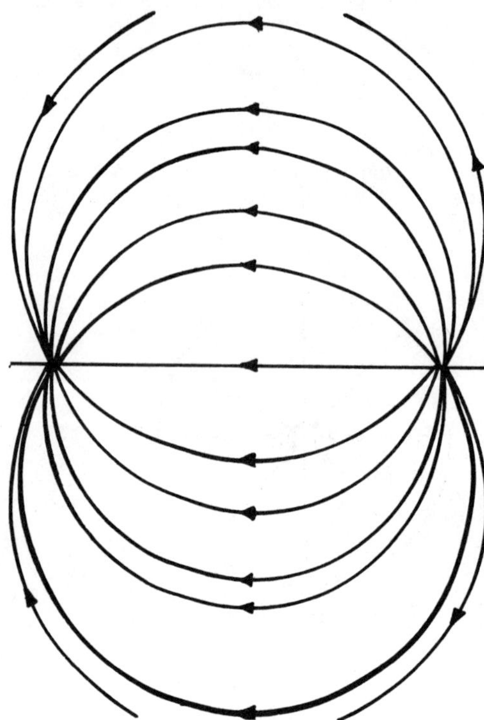

Fig. 3.4. Flow pattern for doublet, equal sink and source.

Combining a source and sink with identical flow forms a doublet. The streamlines for the two-well system are shown in Fig. 3.4. The well pair streamlines are the dominant flow pattern considered. For every gallon of water withdrawn from one well, a gallon of water is injected in the second well. The net withdrawal of water for the system is zero. Larger systems will consist of combined pairs of wells. For potential flow, the effect of each well, sink or source, can be added independently at any point. A natural flow field may also be imposed on the well system. The natural flow is the normal groundwater flow in the aquifer without outside interference. This is the natural flow of water. Figure 3.5 through Fig. 3.7 show an imposed natural flow velocity from different directions imposed on a doublet (12). The magnitude of the natural flow shown in these figures is identical to the flow rate halfway between the sink and source without natural flow. Normally, natural flow is much less than the flow value utilized and can be neglected. In a number of computer programs run for real systems, the natural flow effect could not be detected in the plots. For normal aquifers with low flow rates and minimum distances between wells, the natural flow is usually negligible. Aquifer systems with appreciable flows must be given special independent analysis.

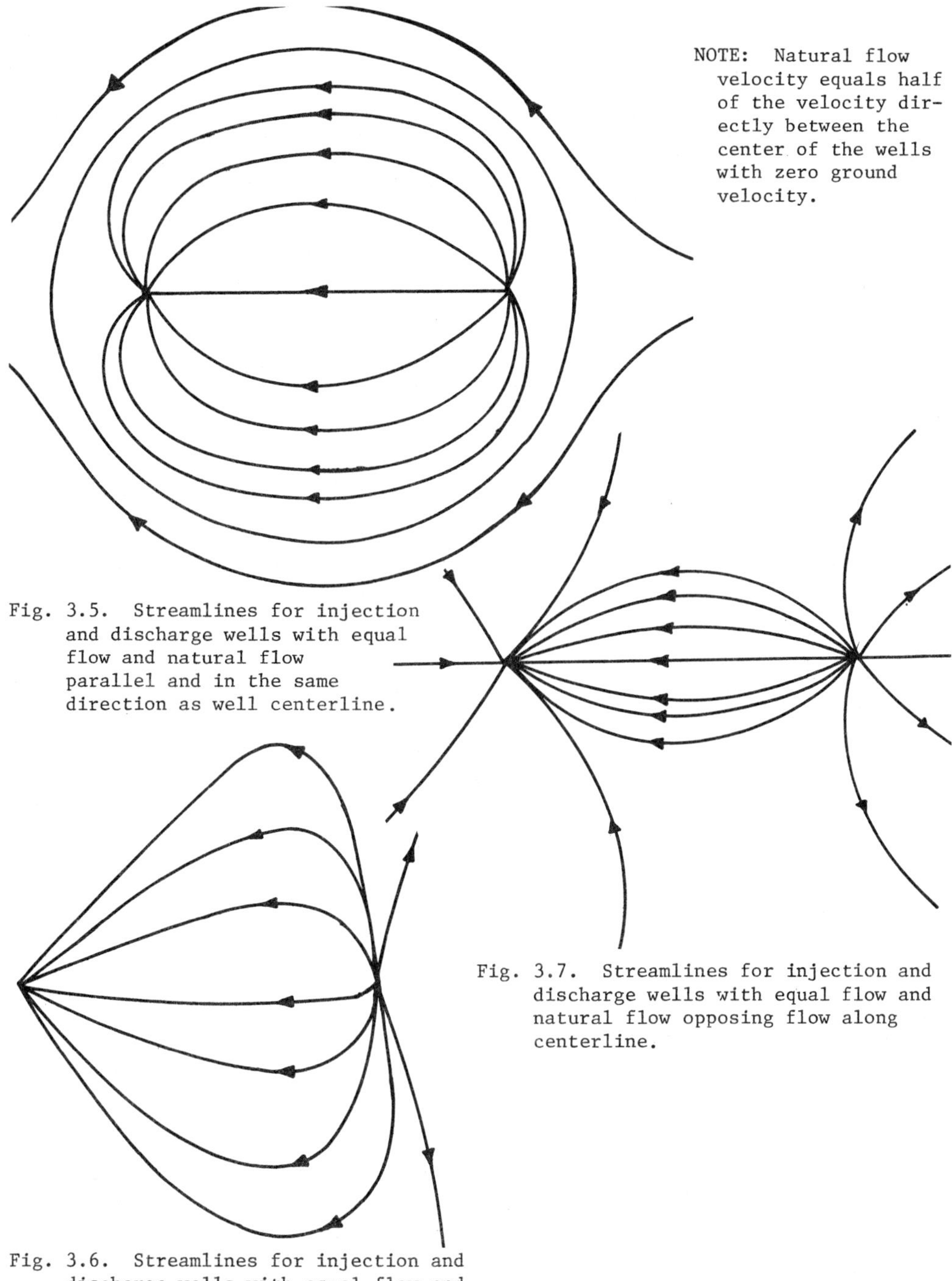

Fig. 3.5. Streamlines for injection
and discharge wells with equal
flow and natural flow
parallel and in the same
direction as well centerline.

Fig. 3.7. Streamlines for injection and
discharge wells with equal flow and
natural flow opposing flow along
centerline.

Fig. 3.6. Streamlines for injection and
discharge wells with equal flow and
natural flow perpendicular to well
centerline.

Well System

A basic well is a cylindrical hole cut from the ground surface through the aquifer as shown in Fig. 3.8.   In most cases a well casing extends from the surface to the aquifer. The casing is commonly stainless steel, cast iron, or PVC (Polyvinyl Chloride).   The annulus between the hole and the casing is sealed so surface water cannot reach the aquifer. The section of the casing set in the aquifer is perforated and normally called "sand screen." The sand screen material is usually stainless steel or PVC. Stainless steel sand screen has the appearance of an ordinary window screen. PVC "screen" consists of very thin slots cut into the pipe. The screen hole sizes or slots vary and are sized with respect to grain size of the aquifer.

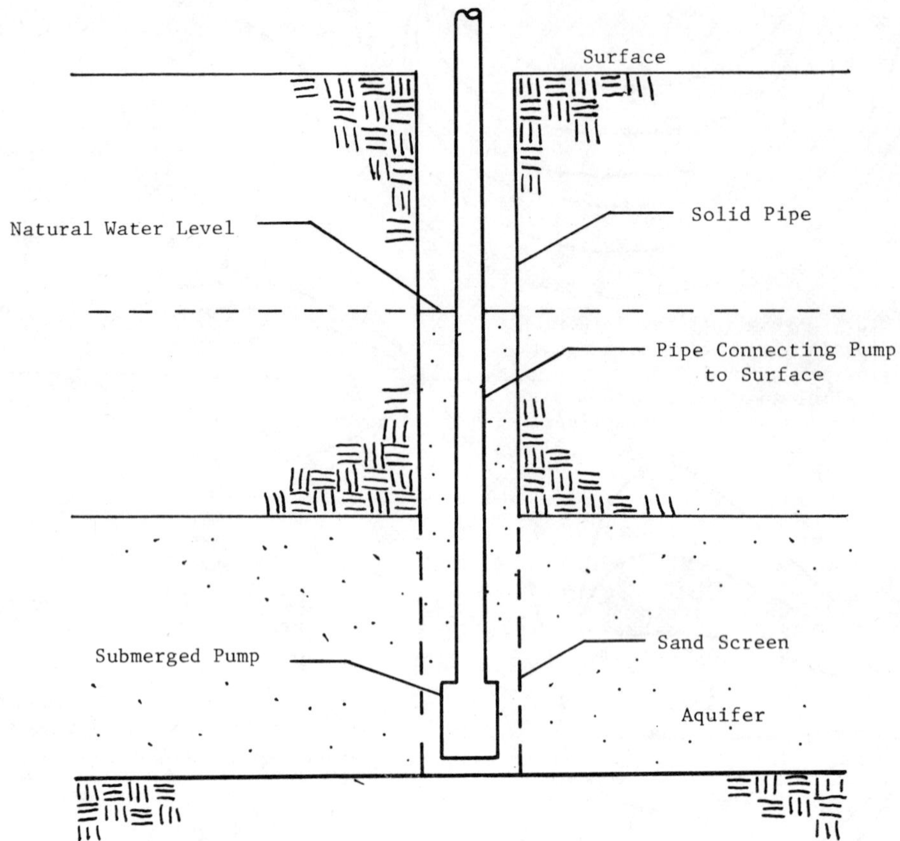

Fig. 3.8. Typical well configuration.

In cases where aquifers consist of hard rock, the well casing sometimes only extends to the hard rock. A seal is made between the casing and rock. No sand screen is required in many hard rock applications.

Submerged below the water level is a pump, generally a submersible or ejector type to lift water to the surface. A pipe connects the pump to the surface distribution system. Pumping systems are discussed in Chapters VI.

Storage Volume

When water is injected into the ground in a single well, it utilizes a large volume of pore space. Water already in the voids must be moved to other voids, resulting in an increase in the volume of stored water. For the well-pair system a transfer between wells minimizes this phenomenon.

For single-well injection the water ideally forms a cylinder around the injection well equal to the height of the aquifer. The volume of this cylinder is equal to the volume of water injected divided by the porosity. This equals:

$$Q/\phi = \pi r^2 b \qquad\qquad (3\text{-}19)$$

where b is equal to aquifer thickness and well screen height and r is the distance from the well. The primary problem with a single-well system is that water may be displaced a distance of many feet by the injected water. This means a very large volume and large amount of aquifer storage area are utilized by one well.

A large source of supply for the water which is heated or cooled must be provided. The water must be filtered and treated before injection into the ground. A disposal source for the water rejected after withdrawal and use must be provided. Because of the large volume of water moved, pressure losses are larger than for a well-pair system.

In the well pair system, water moves between the wells. When one gallon of water is withdrawn, one gallon of water is reinjected. The pattern for the well pair is defined and finite. It will be shown in the next chapter that, in thermal storage, water cycles a number of times before the system reaches its thermal capacity. The pressure loss is limited to the finite distance between wells. For these reasons, each of which will be amplified later, well pairs are used extensively in the systems analyzed.

The theoretical time for the water front to move between wells, asssuming 100 percent porosity, can be calculated by direct integration of the velocity along the line between wells. This velocity, using the notation in Fig. 3.9, is the velocity component from the source plus the velocity component from the sink or:

$$V = \frac{q}{2\pi(r-r_o)} - \frac{q}{2\pi(r+r_o)}$$

$$= \frac{q}{2\pi}\frac{2r_o}{r_o^2-r^2} = \frac{q}{\pi}\frac{r_o}{r_o^2-r^2} \qquad\qquad (3\text{-}20)$$

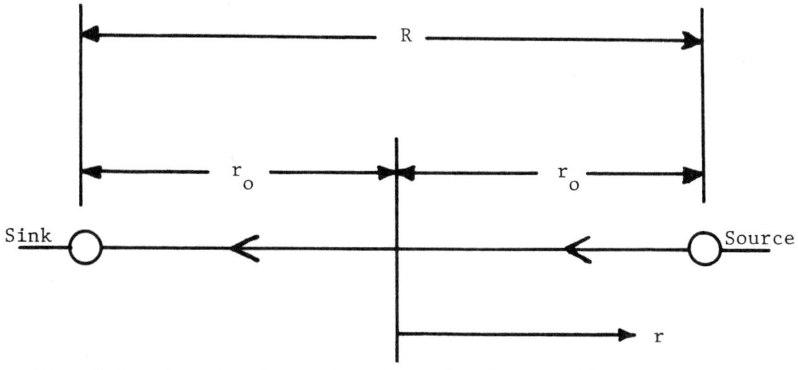

Fig. 3.9. Coordinate system for time integration.

38

Integrating the velocity along the centerline gives:

$$t = \int_{-r_o}^{+r_o} \frac{dt}{V} = \frac{4\pi r_o^2}{3q} = \frac{\pi R^2}{3\,q} \qquad (3\text{-}21)$$

For porosity less than 100 percent a ϕ term must be included. In this case the time based on continuity is:

$$t = \frac{\pi R^2}{3q/\phi} \qquad (3\text{-}22)$$

The time for the passage of the water front between a single well pair is calculated and plotted in Fig. 3.10. Due to the shape of the curves and the limited time ranges, accuracy in reading times is difficult. However, the curves do represent approximate numbers which can be refined with direct calculations. The volume of injected water can be calculated by multiplying the time required for the water front to pass between the wells by the corrected flow rate, thus the volume occupied per unit thickness of aquifer is:

$$v_{water} = \frac{\pi R^2 \phi}{3} = 1.05\ R^2 \phi \qquad (3\text{-}23)$$

The volume of aquifer per unit volume occupied by the injected water is the above value divided by ϕ or:

$$v_{aquifer} = 1.05\ R^2 \qquad (3\text{-}24)$$

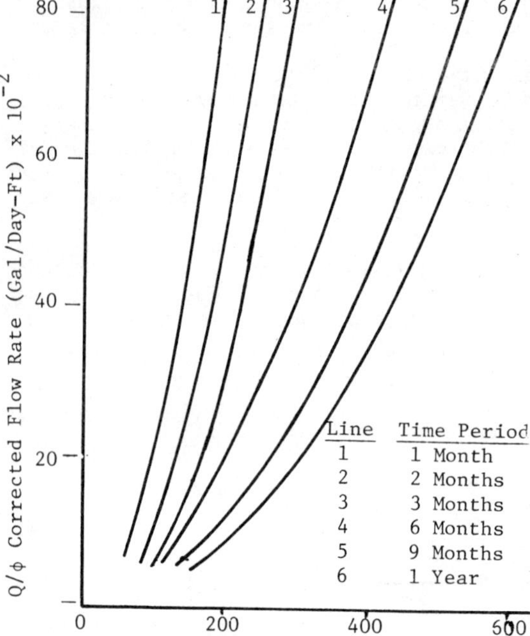

Fig. 3.10. Time required for water front to pass between a single pair of wells.

The volume represents the area per unit thickness between the two wells which is occupied by water when the original water front from the source reaches the sink. This area, as derived by a computer program, is shown in Fig. 3.11. An almost identical pattern is derived theoretically by Collins (4). This is the location of the stored water or energy for the two well system in the horizontal plane.

The identical flow pattern is expected to emerge if the water is injected in increments at various flow rates, which may be positive or negative. The momentum changes in the water flow rates will create variations in the pressure distribution and resulting flow patterns. These variations are assumed to average and produce the storage patterns designated.

The basic computer program derived to produce the outline for the two-well system was developed to include numerous sinks, sources, and natural flow. The horizontal storage area outlines for six wells and ten wells are shown in Fig. 3.12 and 3.13. The available storage areas for each of these systems have

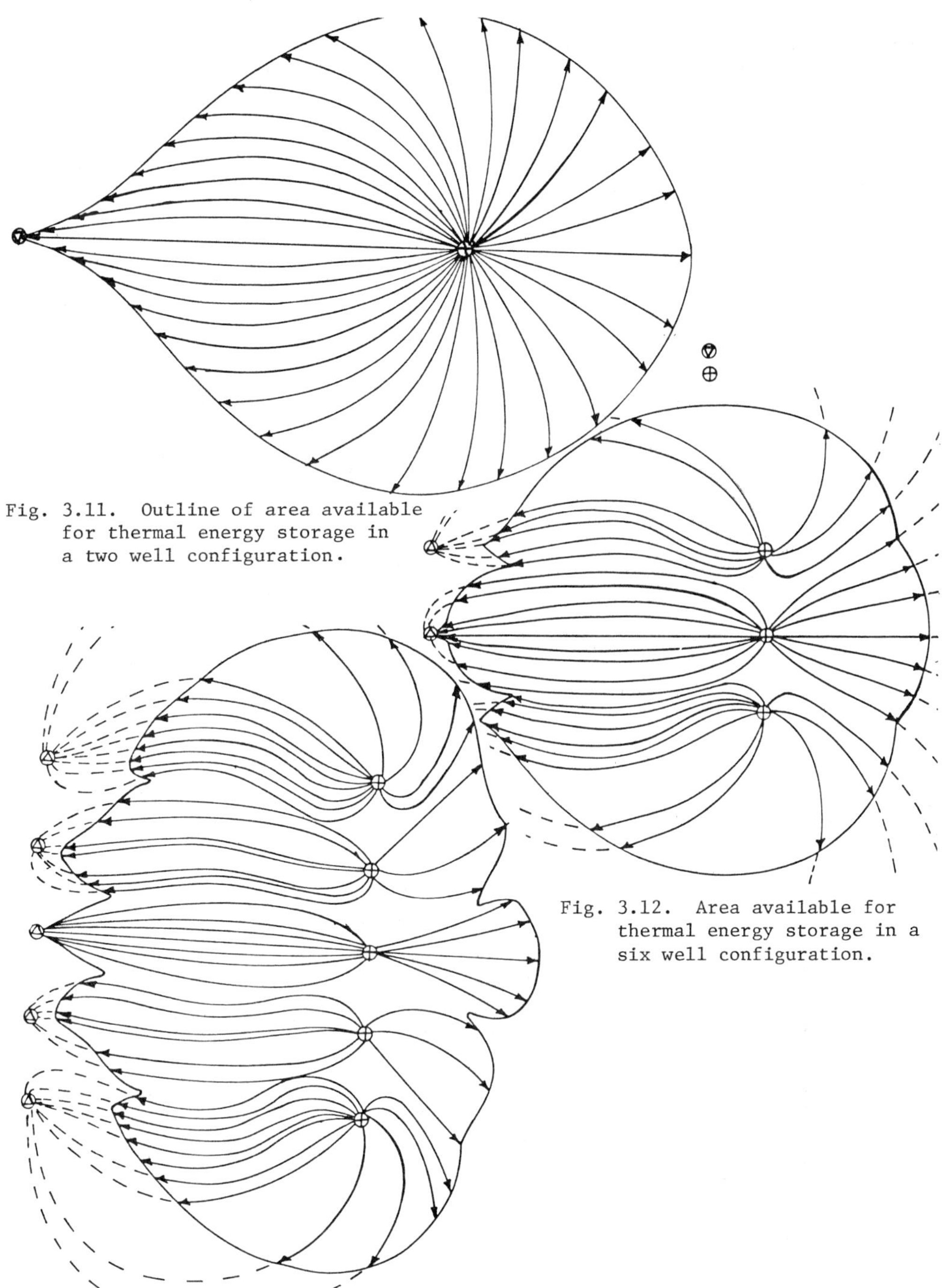

Fig. 3.11. Outline of area available
for thermal energy storage in
a two well configuration.

Fig. 3.12. Area available for
thermal energy storage in a
six well configuration.

Fig. 3.13. Outline of area available for thermal
energy storage in a ten well configuration.

been calculated by computer integration as a function of distance between each well pair and distances between well pairs as defined in Fig. 3.14. The horizontal area available for thermal storage per well pair is:

$$A = CRD \qquad (3-25)$$

where R and D are the identical distances and C is the area coefficient shown in Fig. 3.15. The equation is for cases where all well pairs have the same spacing "R" and are in a line with equal distance "D" between well pairs.

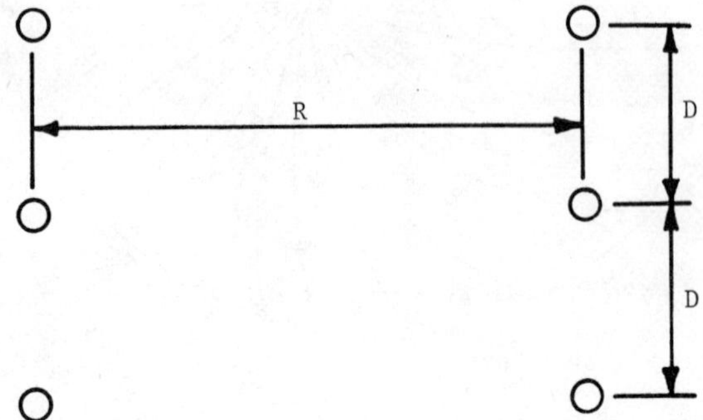

Fig. 3.14. Definition of spacing variables for multi-well system.

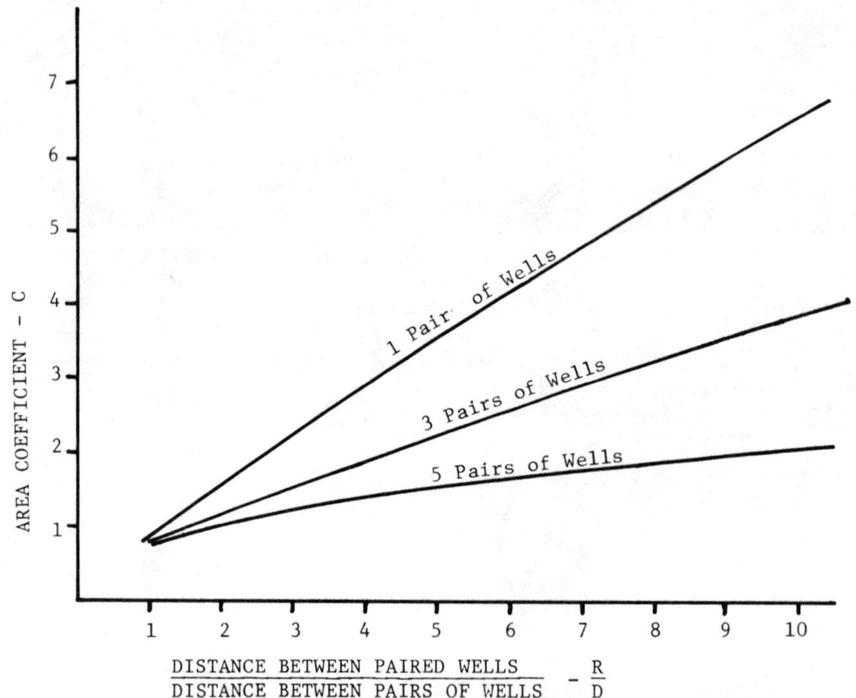

$$\frac{\text{DISTANCE BETWEEN PAIRED WELLS}}{\text{DISTANCE BETWEEN PAIRS OF WELLS}} = \frac{R}{D}$$

Fig. 3.15. Coefficient for effective storage areas.

The relationships developed on water storage will be illustrated by example problems. These relationships are expanded to include energy in the next chapter. For thermal storage, it will be shown that the water can be reused a number of times before the thermal front moves between the wells. A word of caution as to the direct use of the times calculated for water front passage is necessary. The thermal storage area is identical, but the thermal front moves at a slower velocity.

Example: Determine the usable volume in an aquifer 90 ft (b) thick for the following cases:

     a. A single pair of wells 300 ft (R) apart,
     b. Three pairs of wells with 300 ft (R) between wells in each pair and 100 ft (D) between well pairs (Fig. 3.14).
     c. Five pairs of wells with 300 ft (R) between wells in each pair and 100 ft (D) between well pairs (Fig. 3.14).

Solution:

     a. Vol = Thickness times horizontal area
        = $b \cdot 1.05\ R^2$
        = $90\ \text{ft} \cdot 1.05 \cdot (300)^2\ \text{ft}^2$
        = $8.505 \times 10^6\ \text{ft}^3$

     b. Vol_{wp}= $b \cdot CRD$ (C = 1.4 from Fig. 3.15)
        = $90\ \text{ft} \cdot 1.4 \cdot 300\ \text{ft} \cdot 100\text{ft}$
        = $3.78 \times 10^6\ \text{ft}^3/\text{Well pair}$
    Vol_t = $3\ \text{Pairs} \cdot 3.78 \times 10^6\ \text{ft}^3/\text{Pair}$
        = $1.13 \times 10^7\ \text{ft}^3$ (Total Volume)

Note that the three well pairs produce a very small change in storage space, because the velocity between the wells is increased and the large spread of water is lost. Case b could be used to decrease pressure losses experienced with Case a. There is a limit on well output for each well size and drilling three six-inch wells is cheaper than a twelve-inch well in some cases.

     c. Vol_{wp} = $b \cdot CRD$ (C = 1.1 from Fig. 3.15)
        = $90\ \text{ft.} \cdot 1.1\ 300\ \text{ft} \cdot 100\ \text{ft}$
        = $2.97 \times 10^6\ \text{ft}^3/\text{well pair}$
    Vol_t = $5\ \text{well pairs} \times 2.97 \times 10^6 \text{ft}^3/\text{well pair}$
        = $1.485 \times 10^7\ \text{ft}^3$ (Total Volume)

This again is only a minor increase in storage space for the same reasons stated under Case b.

Example: Determine the times required for the water for each combination in the last example to enter the center injection well and reach the center withdrawal well for a flow rate of 400 gpm per injection well. Assume a porosity of 30 percent.

Solution: The water must fill the void in the porous part of the storage volume. Therefore:

$$Qt = \phi\ \text{Vol} \quad \text{or}$$

$$t = \frac{\text{Vol}}{Q/\phi}$$

$$400\ \text{gpm} = 400/7.48\ \text{gal per ft}^3 = 53.5\ \text{ft}^3/\text{min}$$

a. $\quad t = \dfrac{8.505 \times 10^6 \ ft^3}{53.5 \ ft^3/min/0.3}$

$\qquad = 47,660 \ min = 794 \ hr = 33 \ days$

b. $\quad t = \dfrac{3.78 \times 10^6 \ ft^3/ \ well \ pair}{53.5 \ ft^3/min/0.3}$

$\qquad = 21,196 \ min = 353 \ hr = 14.7 \ days$

c. $\quad t = \dfrac{2.97 \times 10^6 \ ft^3}{53.5 \ ft^3/min/0.3}$

$\qquad = 16,654 \ min = 277 \ hr = 11.6 \ days$

This demonstrates the time of water passage between wells calculated from the storage volumes.

Pressure Losses

The pressure losses in the aquifer for withdrawal and injection can be calculated directly for the single well and doublet. The basic equation for flow as a function of permeability can be rearranged for a cylindrical coordinate system:

$$Q = KA \frac{dh}{dl} \quad \text{(Basic Equation 3-9)}$$

$$Q = K2\pi rb \frac{dh}{dl} \quad \text{(Cylindrical Coordinates)} \qquad (3\text{-}26)$$

The pressure loss or head loss for a single well is:

$$\int_0^{\Delta h} dh = \frac{Q}{2\pi Kb} \int_{r_w}^{r_o} \frac{dr}{r}$$

or

$$\Delta h = \frac{Q}{2\pi Kb} \ln \frac{r_o}{r_w} \qquad (3\text{-}27)$$

where r_w is the well radius and r_o is the distance from the well where h is constant. Figure 3.16 shows depression and expansion in the piezometric surface or water table for withdrawal and injection wells respectively. For the single well the effective r_o for a confined aquifer extends to a few hundred feet or to hundreds of miles. As a result the pressure loss can be very large.

The pressure loss for a well pair can be integrated directly with relatively reliable results. Twice the pressure drop for a single well to a distance halfway between the well pair is a good approximation and is used for the well pair. The resulting pressure loss for the well pair is:

$$\Delta h = \frac{Q}{\pi Kb} \ln \frac{r_o}{r} \quad \text{(Single Well Pair)} \qquad (3-28)$$

As can be seen from Fig. 3.16, the relationship is an exponential with the major pressure loss near the well. The head loss is primarily a function of permeability as can be seen from curves in Fig. 3.17. It may be noted that the flow rate shown in the figure is normalized to flow rate per unit of aquifer thickness. As noted previously head loss is independent of porosity.

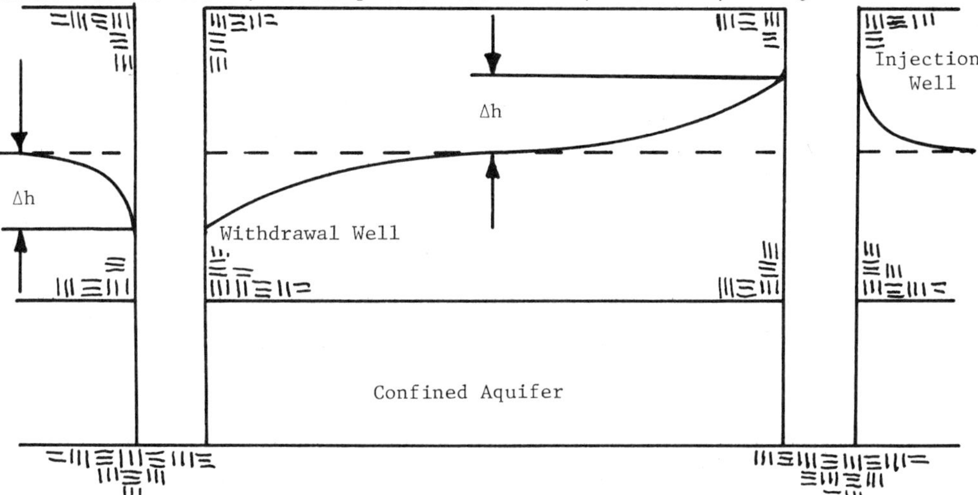

Fig. 3.16. Injection and withdrawal well effect on piezometric surface or water. table.

For multiple pairs of wells with identical separations, pressure loss can be related directly to the times for the water front to move between wells. Pressure loss is directly proportional to velocity. Pressure loss increase is directly proportional to velocity increase. Pressure loss for multiple wells can thereby be directly related to the pressure loss of a single pair of wells and passage times by:

$$\Delta h_{mp} = \Delta h_{sp} \frac{t_{sp}}{t_{mp}} \qquad (3-29)$$

The increased velocity of multiple wells results in an increased head loss.
Pressure loss is the specific weight of water times the head loss or:

$$\Delta p = \gamma \Delta h \qquad (3-30)$$

This development gives a good approximation of the pressure loss for an ideal homogeneous confined aquifer. The pressure losses and/or head from the water table to the heat transfer system are not included in these calcualtions. These are expected to be routine calculations for professionals using aquifers for thermal storage.

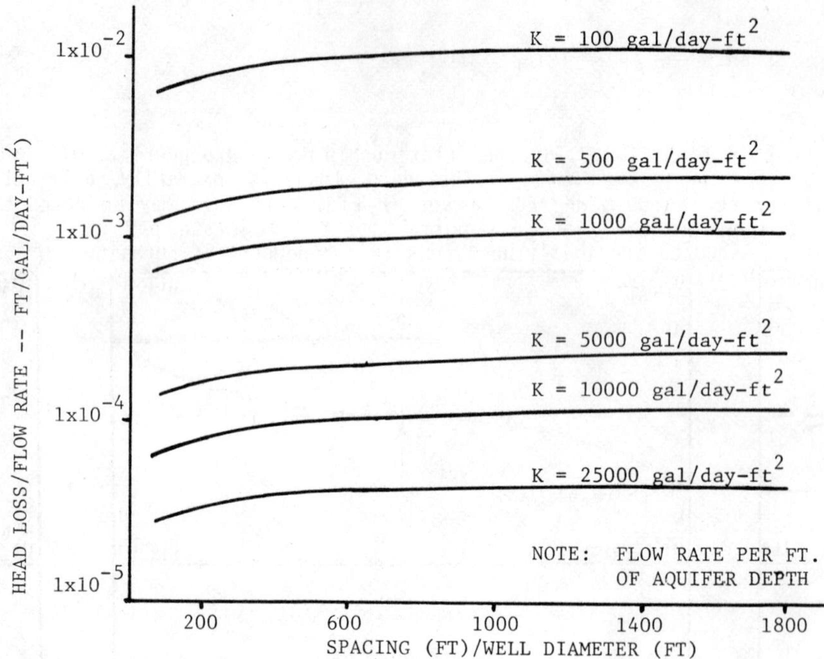

Fig. 3.17. Well pressure drop as a function of spacing and permeability.

Example: For the permeability curve in Fig. 3.17, determine the head and pressure losses for the first case (single well pair) in the previous example. Assume a well diameter of one foot.

Solution: Flow rate equals 400 gpm = 5.76×10^5 gal/day

$$\Delta h = \frac{Q}{\pi Kb} \ln \frac{r_o}{r} \qquad\qquad \frac{r_o}{r_w} = \frac{150 \text{ ft}}{0.5 \text{ ft}} = 300$$

$$\Delta p = \gamma \Delta h \qquad\qquad b = 90 \text{ ft}$$

The values of Δh for the values of K are:

K (gal/day ft^2)	(ft)	(Psi)
25,000	0.46	0.20
10,000	1.16	0.50
5,000	2.32	1.01
1,000	11.6	5.03
500	23.2	10.0
100	116.2	50.4

The pressure loss can also be calculated from the Fig. 3.17. For K equal to 1000 gal/day ft and well spacing to well radius ratio of 300 ft/0.5 ft or 600, the head loss/flow rate equals 1×10^{-3}.

$$\Delta h_{1\ well} = 10^{-3} ft/(gal/day\text{-}ft)\frac{5.76\ (10^{-5}\ gal/day)}{90\ ft}$$

$$= 6.39 \text{ ft for single well}$$

$$\Delta h_{pair} = 2(6.39\ ft) = 12.8\ ft$$

The values for a single injection or withdrawal well in a confined aquifer would normally be twice the above values.

Example: Determine the head and pressure loss in the aquifer for Case b. in the previous examples (3 well pairs) for a K value of 1000 gal/day ft^2.

Solution:

$$\Delta h_{3\ pairs} = \frac{t_{1\ pair}}{t_{3\ pairs}}\ \Delta h_{1\ pair}$$

$$= 13\ ft\ \frac{33\ days}{14.7\ days}\ \text{(times from previous example)}$$

$$= 29.2\ ft$$

$$\Delta p = \Delta h \cdot \frac{62.4\ lb/ft^3}{144\ in^2/ft^2}$$

$$= 12.6\ Psi$$

Example: Determine the head and pressure loss in the aquifer for Case c in the previous example (5 well pairs) for a K value of 1000 gal/day ft^2.

Solution:

$$\Delta h_{5\ pairs} = \frac{t_{1\ pair}}{t_{5\ pairs}} = \Delta h_{1\ pair}$$

$$= 13\ ft\ \frac{33\ days}{11.6\ days}\ \text{(times from previous example)}$$

$$= 37.0\ ft = 16.0\ Psi$$

The values for head loss or pressure loss include only the losses between wells. The head to the surface and pipe head losses are not included.

Summary

The water flow patterns have been developed for well combinations in a homogeneous confined aquifer. Calculations for the times for transfer of the water front and the pressure (head) loss between wells have been developed into a form useful for design work. Proper application of these mathematical developments will provide reasonable design values. Some safety factor should be applied. This factor will vary with knowledge of the aquifer. The factor for pressure loss should normally be higher than the factors for storage volume and transit times.

Additional engineering approximations will be required for non-homogeneous aquifers of varying thickness and other properties. An additional problem which must be accounted for is the natural flow in aquifers. Since this applies mainly to thermal storage, an analysis is contained in the next chapter. Only aquifers with low natural flows can be utilized.

These relationships have been developed for a confined aquifer. They are also good approximations for aquifers confined only at the bottom. The primary variation will be the thermal energy mixing due to bouyancy effects. This is discussed in more detail in the next chapter.

References

1. Todd, D.K., Groundwater Hydrology, Wiley, New York, 1963.

2. Craft, B.C. and Hawkins, M.F., Applied Petroleum Reservoir Engineering, Prentice-Hall, Englewood Cliffs, New Jersey, (1959).

3. Darcy, H., Les fontaines publiques de la ville de Dijon, Dalmont, Paris, (1856).

4. Collins, R.E., Flow of Fluids Through Porous Materials, Reinhold, New York, (1961).

5. De Wiest, R.J., Geohydrology, Wiley, New York, (1965).

6. Pirson, S.J., Oil Reservoir Engineering, McGraw-Hill, New York, (1958).

7. Scheidegger, A.E., The Physics of Flow Through Porous Media, MacMillan, New York, (1960).

8. Clark, S.P., Handbook of Physical Constants, The Geological Society of America, (1966).

9. Hunsaker, J.C., and Rightmire, B.G., Engineering Applications of Fluid Mechanics, McGraw Hill, New York, (1947).

10. Rauscher, M., Introduction to Aeronautical Dynamics, John Wiley, New York (1953).

11. Henry, H.R., McDonald, J.R., and Alverson, R.M., Aquifer Performance Tests under Two-Phase Flow Conditions, BER Report 131-118, The University of Alabama, (March 1971).

12. Moly, F.J., Warman, J.C., and Thomas, E.J., Aquifer Storage of Heated Water: Experimental Study, Auburn University, (1978).

13. Schaetzle, W.J., Brett, C.E., and Seppanen, M.S., Heat Pump Centered Integrated Community Energy Systems, NTIS No. ANL/CNSV-TM-25, (August 1979).

CHAPTER IV

THERMAL ENERGY STORAGE IN AQUIFERS

Thermal energy is stored in the rock and water which comprise the aquifer volume. Energy is injected and withdrawn after storage in the aquifer. As water flows through the aquifer, it relinquishes energy to the rock and water in the aquifer storage volume. A thermal energy front thereby follows the water front at a slower velocity. This is the major concept which is developed and analyzed in this chapter.

Energy losses from the aquifer occur in a number of ways. The conduction to the confining layers and to the aquifer volume ahead of the thermal front occur in all aquifers. Some convection to the aquifer in front of the thermal front also occurs. Some free convection thermal mixing due to buoyancy instabilities degrades the energy. This is residual energy which is lost to the aquifer. Another loss occurs through the natural flow in the aquifer. This loss is a function of the natural flow velocity and aquifer porosity.

The losses which occur are independent of whether the storage water is cold or hot. The loss phenomena are identical for a heat sink or heat source. The difference between storage temperature and natural aquifer temperature affects the magnitude of loss but has a small effect on percentage loss if natural aquifer temperature is considered as a base temperature.

Aquifer thermal energy storage deals with very large storage systems provided by nature. The surface area of the storage volume is small per unit volume compared to man-made storage. In addition, the Earth is a good insulator; frequently several hundred feet of rock may overlie the aquifer. Very little if any change in temperature can be detected sixteen feet below the ground's surface throughout an annual cycle in most areas of the world. Natural spring or well water is essentially the same temperature the entire year, and air in major underground caves is essentially at a constant temperature the entire year. Aquifers provide a major renewable energy storage resource available for thermal energy storage applications.

Energy Storage Capacitance

Thermal energy is stored in both the water and rock in each unit volume of aquifer. Storage is a function of the porosity, and other water and rock physical and thermal properties. The thermal energy capacity per unit volume of aquifer per degree of temperature change is the thermal capacity in the rock plus the thermal capacity in the water. Thermal capacity is:

48

$$q = (c_p \rho)_{rock} (1-\phi) + (c_p \rho)_{water} \phi \qquad\qquad (4-1)$$

where c_p is the specific heat and ρ is the specific density. The total heat capacity per unit volume is the specific capacity times the temperature variation. The specific heat and density are given in Table 4.1 for various rock types. The specific heat and density of water are 1 Btu/lb°F and 62.4 lb/ft³ respectively.

TABLE 4.1   Thermal Properties of Rock*

Substance	Specific Gravity	Specific Heat Btu/lb°F
Granite	2.6-2.7	0.20
Limestone	2.1-2.86	0.217
Marble	2.5-2.86	0.21
Sandstone	2.0-2.6	0.22
Earth	1.2-1.8	Varies with moisture
Sand and Gravel	1.4-2.2	0.19
Quartz	2.5-2.6	0.17-0.28
Coal	1.1-1.8	0.3
Graphite	1.6-2.7	0.165 to 0.390

*Values compiled from Ref. 1 and 2

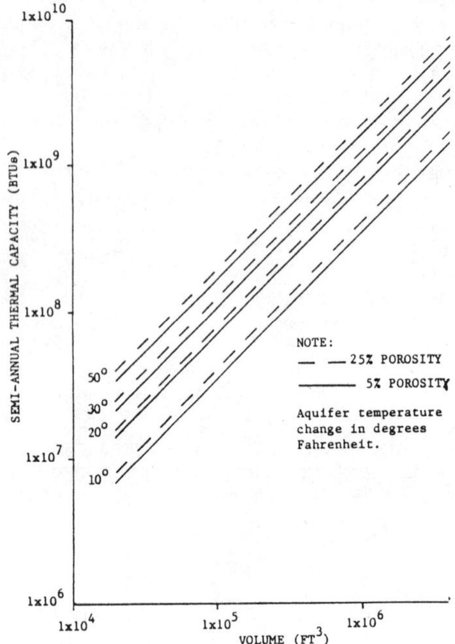

Fig. 4.1. Aquifer thermal energy storage capacity.

With a specific heat of 0.20 Btu/lb°F, and a specific gravity of 2.6 for rock, and an aquifer porosity of 0.2, the specific thermal capacitance per unit volume of aquifer is 38 Btu/ft³°F (1400 kJ/m³). At this value, a rock volume 100 ft by 200 ft by 100 ft (30m by 60m by 30m) has a storage capacity of 1.5 x 10⁹ Btu (1.4 x 10⁹ kJ) with a 20°F (11°C) temperature difference. With a 200°F (111°C) temperature difference, the thermal storage capacity is 1.5 x 10¹⁰ Btu (1.4 x 10¹⁰ kJ). Figure 4.1 gives the variation of thermal capacity as a function of porosity for a number of temperature differences. To determine the thermal energy storage for large temperature differences, the temperature difference is multiplied by ten to increase the thermal capacity by ten. As may be seen, the effect of porosity is minor. The required volume of storage can be calculated from the capacity numbers. The ideal volume required to store energy Q is:

$$V = \frac{Q}{q \Delta T} \qquad\qquad (4-2)$$

This volume must be corrected for thermal storage efficiency, natural aquifer flow losses, and a safety factor.

Example: Determine the ideal storage volume required to store 10^9 Btu in an aquifer with a porosity of 0.3 at 100°F in an aquifer at 60°F.

Solution: The specific thermal capacity is:

$$q = (c_p \rho)_{rock} (1 - \phi) + (c_p \rho)_{water} \phi$$

For a rock specific heat of 0.20 Btu/lb°F and specific gravity of 2.6:

$$q = 0.20(2.6)\ 62.4(1-0.3) + 1.0(62.4)(0.3)$$
$$= 41.4\ Btu/ft^3\ °F$$

The volume required is:

$$V = \frac{Q}{q\Delta T} = \frac{10^9}{41.4(100-60)} = 6.04 \times 10^5\ ft^3$$

Thermal Front

Thermal energy storage in the aquifer is a combination of the energy stored in the rock and water in the pore volume. As water moves through the aquifer, thermal energy is transferred to the rock. Both the rocks and water in the pore volume contain thermal energy. As a result, the thermal front moves through the aquifer at a lower velocity than that of the water front. Since the thermal energy must be shared, its motion is slower than the motion of the water.

This movement results in a thermal front following the water front through the aquifer. The thermal energy storage concept is based on this model.

A similar common phenomenon is the thermal front movement through water pipes. When a hot-water spigot is opened, the water remains cold for a seemingly long period. As the water begins to warm, it becomes hot almost instantly. Physically the hot water in the hot water tank begins moving towards the spigot as it is opened but loses part of its thermal energy to the water pipe. This water which is cooled by the pipe arrives at the spigot before the water becomes warm. The hot water continuously loses its thermal energy to the pipe until the pipe reaches the hot water temperature. As a result, two waves move through the water line. The first is the water front which moves with water velocity. The second is a thermal front which moves at a lower velocity behind the water front. The water front must fill only the pipe volume but the heat energy must fill the pipe volume and also heat the pipe. The thermal front is therefore slower. The phenomenon arises because the heat transfer (forced convection) to the water pipe is much larger than the conduction along the water route, the conduction along the pipe, or the free convection from the outside of the water pipe. The temperature of the pipe at any point becomes approximately the temperature of the water at that point.

The situation in an aquifer is similar but much more clearly defined. The convection between water and aquifer is an order of magnitude higher than in the water pipe. This is due to a large convection heat transfer resulting from the large amount of surface area per unit volume. If spheres are packed in a cubic arrangement (Fig. 4.2), each sphere ($\phi = 0.476$), (3) occupies a volume equal to its diameter cubed. For 0.001-ft diameter sand particles, there are 10^9 particles per cubic foot. The surface area to unit volume ratio of the particles equals 785 ft^2/ft^3. When the rhombohedral model (Fig. 4.2) is considered, the packing ($\phi = 0.26$), (3) is denser and surface area increases to over 1400 ft^2/ft^3.

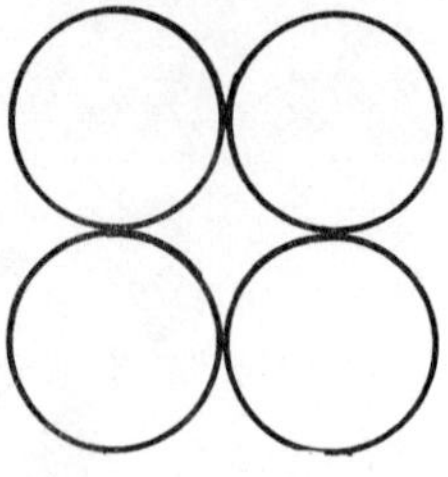

Cubic Arrangement

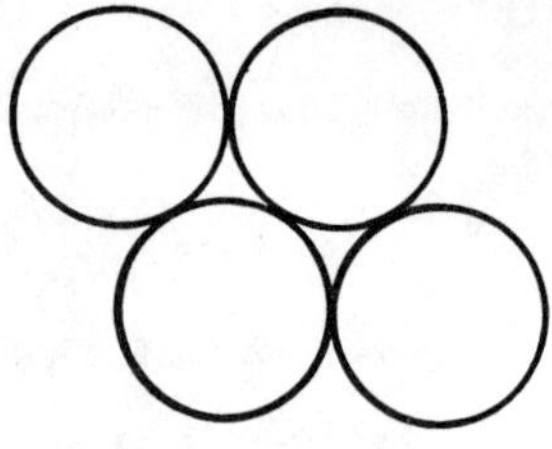

Rhombohedral Arrangement

Fig. 4.2. Models for packing spheres..

Due to the large area to volume ratio the internal convection heat transfer ($Q = hA \Delta T$)   between water and aquifer completely dominates the conduction. Basically the water temperature and rock temperature at any location are identical.

The water front velocity and thermal front velocity concept will be derived mathematically for both the water pipe and the aquifer. The pipe model and the aquifer model will be developed in parallel sequence. This is done to help explain the temperature front movement relative to the water front movement. The basic assumption in both cases is that the temperature of the pipe and the rock comprising the aquifer, is identical to the temperature of the water at any location. The mathematical development is based on the law of the conservation of energy as applied to a small volume element of pipe and water and then rock and water (Fig. 4.3). The conduction of energy to the surroundings of these elements is assumed negligible for the basic derivation.

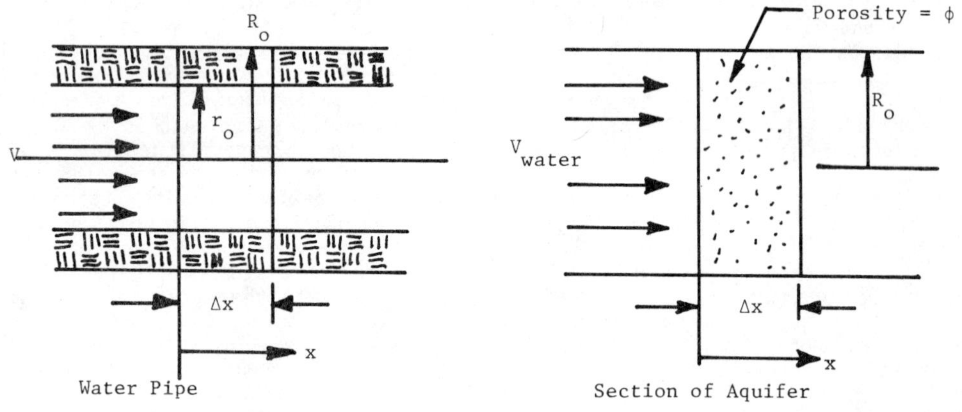

Fig. 4.3. Mathematical models for water and thermal fronts.

Applying the law of conservation of energy to the pipe element, or the energy in (by mass transfer) minus the energy out (by mass transfer) equals the energy stored:

$$E_{in} = \text{water mass flow} \cdot \text{water specific heat} \cdot \text{time increment} \cdot \text{temperature at location x}$$

$$= (V\rho A_i)_{water} c_{p_{water}} \Delta t\ T_x$$

$$E_{out} = \text{water mass flow} \cdot \text{water specific heat} \cdot \text{time increment} \cdot \text{temperature at location } x + \Delta x$$

$$= (V\rho A_i)_{water} c_{p_{water}} \Delta t\ T_{x+\Delta x}$$

$$E_{stored} = \text{thermal capacitance of water plus pipe times the temperature change}$$

$$= (\rho A_i C_p)_{water} + (\rho A_p c_p)_{pipe} \Delta x (T_{t+\Delta t} - T_t)$$

Applying conservation of energy:

$$E_{in} - E_{out} = E_{stored}$$

$$(V\rho A_i c_p)_{water} \Delta t (T_x - T_{x+\Delta x}) = (\rho A_i c_p)_{water} + (\rho A_p c_p)_{pipe} \Delta x (T_{t+\Delta t} - T_t) \qquad (4\text{-}3)$$

Dividing both sides by $\Delta x \Delta t$ and taking the limit as Δx and Δt approach zero results in the following partial differential equation:

$$(V\rho A_i c_p)_{water} \frac{\partial T}{\partial x} = ((\rho A_i c_p)_{water} + (\rho A_p c_p)_{pipe}) \frac{\partial T}{\partial t} \qquad (4\text{-}4)$$

Since $T(x)$ and $T(t)$ must be independent of each other, both sides of the above equation must equal a constant. In this case the constant must equal zero or upon integrating:

$$T(x) = constant \qquad (4\text{-}5)$$

and

$$T(t) = constant \qquad (4\text{-}6)$$

The initial condition is $t = 0$, $T = T_0$ or the initial temperature of the pipe and contained water before the hot water flow is commenced. The final condition is for a very large t, $T = T_f$ or the temperature of the hot water which enters the pipe. This assumes the heat transfer from the water pipe to the surroundings is negligible. As the water flows, a water front representing the water interface entering the pipe at $t = 0$ moves through the pipe at velocity V. A thermal front moves through the pipe behind the water front. The temperature of pipe and water in front of the thermal front is T_0 and the temperature of the pipe behind the thermal front is T_f. The location of the thermal front is determined by the law of conservation of energy. The energy into the pipe has raised the temperature of a section of pipe and water to temperature T_f. The total mathematical solution of differential equation 4-4 is a square wave where:

$$T = T_o \text{ for } x > \frac{(\rho A_i c_p)_{water} V}{(\rho A_i c_p)_{water} + (\rho A_p c_p)_{pipe}} t \qquad (4\text{-}7)$$

$$T = T_f \text{ for } x < \frac{(\rho A_i c_p)_{water} V}{(\rho A_i c_p)_{water} + (\rho A_p c_p)_{pipe}} t \qquad (4\text{-}8)$$

The term relating to x represents the location of the thermal front at any time t. Physically the water front moves through the pipe with the thermal front lagging the water front as a fraction of the energy in the water must be shared with the pipe to raise it to the hot water temperature. In reality thermal energy losses occur from the pipe, conductivity takes place parallel to the pipe in both the pipe and water, and the water and pipe are not exactly at the same temperature. However, this result is a good approximation. The thermal front velocity to water front velocity is:

$$V_{thermal} = \frac{(\rho c_p A_i)_{water}}{(\rho c_p A_i)_{water} + (\rho c_p A_p)_{pipe}} \cdot V_{water} \qquad (4-9)$$

The time for the thermal front to the time for the water front to reach any location is related as:

$$t_{thermal} = \frac{(\rho c_p A_i)_{water} + (\rho c_p A_i)_{pipe}}{(\rho c_p A_i)_{water}} \, t_{water} \qquad (4-10)$$

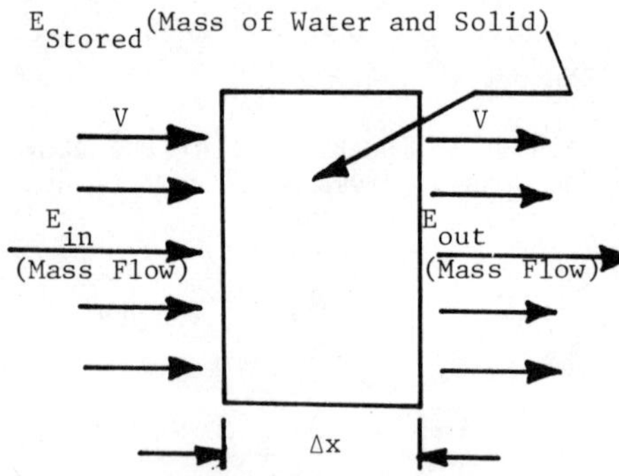

The above calculations are for hot water entering a cold pipe; however, the results are identical for cold water entering a hot pipe. When the cold water spigot is turned on the water out is at room temperature for a period and then the water seems to turn cold instantly.

The derivation for the aquifer is identical except that the assumption where the temperature of the water and aquifer at any location is identical is much more exact than for the pipe and water. Again, the conduction into and out of the element is neglected. The heat transfer to and from the element is by the thermal energy in the water entering and leaving the element.

The energy balance of the element in Fig. 4.4 has the following energy changes which are very similar to the pipe example:

Fig. 4.4. Element for aquifer energy balance.

$$E_{in} = V\rho\phi Ac_p \Delta t T_x \quad \text{(for water flow)}$$

$$E_{out} = V\rho\phi Ac_p \Delta t T_{x+\Delta x} \quad \text{(for water flow)}$$

$$E_{stored} = ((\rho c_p \phi)_{water} + (\rho c_p (1 - \phi))_{rock}) A\Delta x (T_{t+\Delta t} - T_t)$$

Here ϕ represents the fraction of volume in the element which is void and filled with water. The term $(1-\phi)$ represents the fraction of volume which is solid rock. It is also assumed that the fraction of area element A through which water flows is ϕ . This is not necessary as this flow of water is normally given as volume flow per unit area. The term in brackets represents the thermal capacity per unit volume of aquifer which, multiplied by the volume $A\Delta x$ and by T, gives the thermal energy in the volume element. Applying conservation of energy results in the following equation:

$$E_{in} \quad - \quad E_{out} \quad = \quad E_{stored}$$

or

$$V\rho\phi Ac_p\Delta t \ (T_x - T_{x+\Delta x}) = ((\rho c_p\phi)_{water} + (\rho c_p(1-\phi))_{rock})A\Delta x \ (T_{t+\Delta t}) \qquad (4-11)$$

Dividing by $A\Delta x\Delta t$ and taking the limit as Δx and Δt approach zero, results with the differential equation:

$$(V\rho c_p)_{water} \ \frac{\partial T}{\partial x} = ((\rho c_p\phi)_{water} + (\rho c_p(1-\phi))_{rock}) \ \frac{\partial T}{\partial t} \qquad (4-12)$$

The solution must consist of two parts. Temperature must be a function of x and also temperature must be a function of t, independent of each other. Since T (x) and T (t) must be independent of each other both equations must equal a constant. In this case the constant equals zero as with the pipe case or:

$$T(x) = constant \qquad (4-13)$$

$$T(t) = constant \qquad (4-14)$$

It is assumed that the entire aquifer is at the initial condition T_0. After an infinite amount of time with water injection the aquifer must reach the incoming water temperature T_f. The mathematical boundary conditions are:

$$t < 0 \qquad T = T_o$$
$$t \rightarrow \infty \qquad T = T_f$$

As the water flows through the aquifer, the energy supply due to the temperature difference is transferred to the rock. The final energy in a volume is the energy in the rock plus the energy in the water. As a result the heating of the aquifer is much slower than the movement of the water. A temperature front follows the water front, at a lower velocity. The solution to the differential equation is:

$$T = T_o \ for \ x > \frac{(\rho c_p\phi)_{water}}{(\rho c_p\phi)_{water} + (\rho c_p(1-\phi))_{rock}} \ Vt \qquad (4-15)$$

$$T = T_f \ for \ x < \frac{(\rho c_p\phi)_{water}}{(\rho c_p\phi)_{water} + (\rho c_p(1-\phi))_{rock}} \ Vt \qquad (4-16)$$

The term Vt is the distance that the water front has moved whereas x is the distance the thermal front has moved. These distances in a system where the water does not move through a channel of constant area, radial flow for example, do not have the above ratio. However, a direct relationship exists between the thermal and water front velocity at any point.

The thermal front velocity is related to the water front velocity at any point as:

$$V_{thermal} = \frac{(\rho c_p \phi)_{water}}{(\rho c_p \phi)_{water} + (\rho c_p (1-\phi))_{rock}} \, V_{water} \qquad (4\text{-}17\,)$$

The time for the thermal front to arrive at any location as a function of the water front time for the identical location is:

$$t_{thermal} = \frac{(\rho c_p \phi)_{water} + (\rho c_p (1-\phi))_{rock}}{(\rho c_p \phi)_{water}} \, t_{water} \qquad (4\text{-}18\,)$$

As may be seen the thermal velocity is lower than the water velocity and the time of passage for the thermal front is larger than for the water front. The variation is primarily a function of porosity. The ratios for rock with a specific gravity of 2.6 and a specific heat of 0.22 Btu/lbm^oF and various porosities are given in Table 4.2.

TABLE 4.2 Ratios of Velocity and Time as a Function of Porosity

Porosity	$V_{thermal}/V_{water}$	$t_{thermal}/t_{water}$
0.01	0.017	57.63
0.05	0.084	11.87
0.10	0.162	6.15
0.15	0.236	4.24
0.20	0.304	3.29
0.25	0.368	2.72
0.30	0.429	2.33
0.40	0.538	1.86
0.50	0.637	1.57

The concept in the aquifer heat transfer resulting from this derivation, or the ideal solution is:

1. A water front moves out from the injection well as a function of injection flow rate as noted in Chapter III.

2. A thermal front follows the water front at a lower velocity. The temperature change at the front is basically a step function.

3. When the flow is reversed the thermal front flow is reversed and the injected energy (source or sink) is recovered without deterioration. Ideally there is no deterioration of temperature. The recovery temperature is the injected temperature.

This phenomenon is very similar to the energy variation in a pebble bed heat exchanger, where air passes through small rock fragments. For short duration storage the above equation is accurate. For large storage times the square wave deteriorates somewhat. The thicker the aquifer the closer a square wave solution represents the thermal front. In a 300-ft thick aquifer, the storage can be represented very closely by a square wave. The conduction losses are relatively small. Very little effect can be seen eight feet below or above the aquifer. The deterioration of the thermal front can be compared to the water pipe example representing an aquifer. The hot water becomes warm rapidly but not as a step temperature change.

The derivation has basically represented hot water storage. The solutions for cold storage are identical. Only the direction of heat conduction changes.

Real Energy Storage System

In a real storage system energy losses exist. These losses include conduction losses to the confining layers, conduction losses in the aquifer forward of the thermal front, water mixing losses in the aquifer, and convection losses due to the natural flow in the aquifer. These losses have been analyzed by a number of authors (4,5,6) through aquifer simulation on the computer. The available experimental data are very limited. The data are general and not clearly defined. The following aquifer energy analysis has been developed at The University of Alabama. Cross references and comparisons are made to work by other research teams.

The concept is based on the thermal front moving through the aquifer with conduction losses superimposed on the thermal wave. The computer runs and computer program development are the backbone of Ansari's dissertation (6). In these simulations hot water is injected into the aquifers for different time periods. These times vary from twelve hours through six months. The water is withdrawn for similar periods of time. The time periods match solar energy conditions, overnight storage, several days storage for bad weather, and the annual cycle. The computer program is structured to simulate any real time utilization of the aquifer.

A new concept, the conditioned aquifer, is investigated. Under this concept hot or cold water is injected into the aquifer for a prescribed period. This energy is basically written off. It conditions the aquifer and acts as a buffer or safety factor. For example, 400°F (204°C) water is injected into an aquifer for six months. Then water is injected and recovered over a 30-day period. The recovered energy is withdrawn at a temperature within a few degrees of the injection temperature, and basically 100 percent of the quantity of the short term injected energy is recovered. A utilization of this example could be a solar power plant. The required temperature for the steam cycle is critical. As the power plant is constructed, the energy from the installed collectors is immediately stored. By the time the power plant starts operating, the aquifer is conditioned.

Energy recovery for efficiency evaluations is defined as the energy recovery until the temperature drops by 25 percent between input temperature and original aquifer temperature. This definition follows the recommendation of proposed ASHRAE Standard 94.2P for Thermal Storage Devices with Electrical Input and Thermal Output Based on Thermal Performance. The standard assumes energy is normally recovered by air or water flow through the storage device. This standard is not similar to aquifer storage, but does set a precedent. ASHRAE standard 94-77 suggests a plot of temperature versus time or quantity recovered for thermal evaluation of energy storage devices. This is ideal, but difficult to generate, especially before utilization. An aquifer can not be tested and evaluated as can smaller identical storage devices. Some curves of aquifer computer stimulation of these concepts are included. The 25 percent degradation or 75 percent recovery temperature is recommended as a figure for general recoveries. The actual definition must be derived for each application. If 95 percent of temperature must be recovered to run a steam boiler, 75 percent recovery is meaningless.

Figure 4.5 is a computer run (6) to simulate the water injection of Davison's group at Texas A&M (7). A well pair 1340 ft apart comprises the storage system. Approximately 7.7 million gallons of water were injected into a 15-ft thick aquifer over a three-month period. For the computer simulation the input flow was averaged over this period and injected at 48°F. The original aquifer temperature was 68°F (20°C) and the average injection temperature 48°F (9°C) with variations from 39°F to 53°F (4°C to 12°C). The computer simulation is shown as lines in Fig. 4.5. The circles represent experimental data. Data were obtained only above the aquifer centerline. The isotherms show the conduction heat transfer above and below the aquifer and through the aquifer. Note that the conductivity in the aquifer is approximately twice the value as in

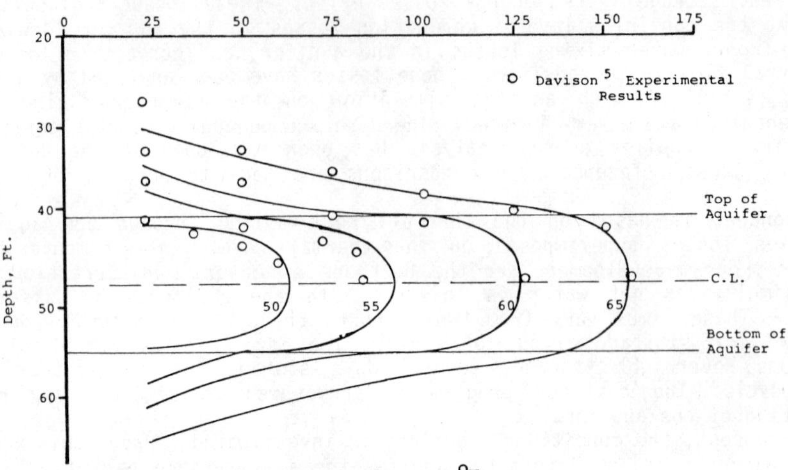

Fig. 4.5. Ground water isotherms, °F.

the confining layers. No buoyancy related mixing can be determined from the experimental results. This study has used cold water injection into the aquifer and the injected water could be slightly below the aquifer centerline. LeCroy (8) found a more pronounced effect using hot water in a laboratory model of an aquifer. Again the injected water (hot in this case) is centered below the aquifer center line.

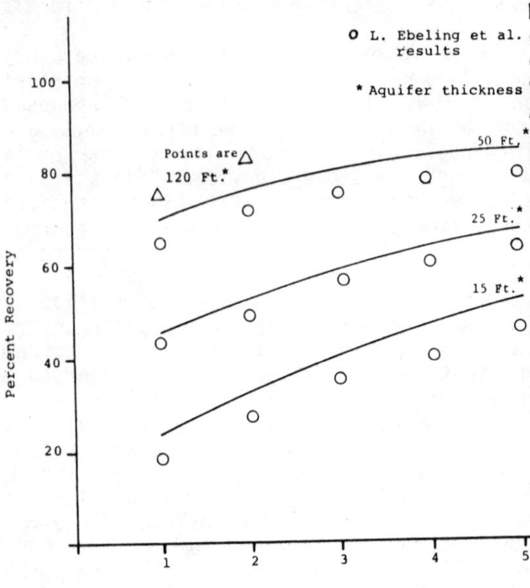

Fig. 4.6. Thermal energy recovery after annual cycles.

Davison (5) also made a number of computer runs to determine the recovery efficiency. The same computer runs were duplicated at The University of Alabama. Figure 4.6 shows recovery efficiencies (6) for different values of aquifer thickness. The curves are for six-month storage cycled annually. Davison's values are shown as circles. Two values for a 120-ft thick aquifer are also shown. The difference between the programs is the method by which the mass heat transfer is handled. The recovery values are considered reasonable. Some actual experimental data for a well defined aquifer are required. The predicted 20 percent recovery at Texas A & M resulted for one cycle of cold storage. LeCroy (8) has cycled water in the model aquifer for short periods of time. The model aquifer is only two ft thick, however a 20 percent recovery resulted on the first cycle. The increase in performance for the second cycle duplicated the predicted increase in performance by Davison (5) and Ansari (6).

A number of aquifer storage simulation computer runs were made (6). These are for conditioned and unconditioned aquifers 120 ft thick. The two wells are spaced 400 ft apart. A basic injection and withdrawal rate of 84 gpm is utilized. The runs were for:

   a. Twenty four hour cycle simulating overnight storage.
   b. Several day (eight) cycle simulating a bad weather period, and
   c. Twelve month cycle simulating the annual variation.

Emphasis has been placed on centerline temperatures to show the residual energy left in the aquifer. Figure 4.5 show the effect to the confining layers. At a depth greater than sixteen ft (arbitrary) little effect is noted. The system in Fig. 4.5 can basically be split with a section added to the center of the aquifer for thicker aquifers. The following figures for the 120-ft thick aquifer are included to give an insight into aquifer energy storage recovery.

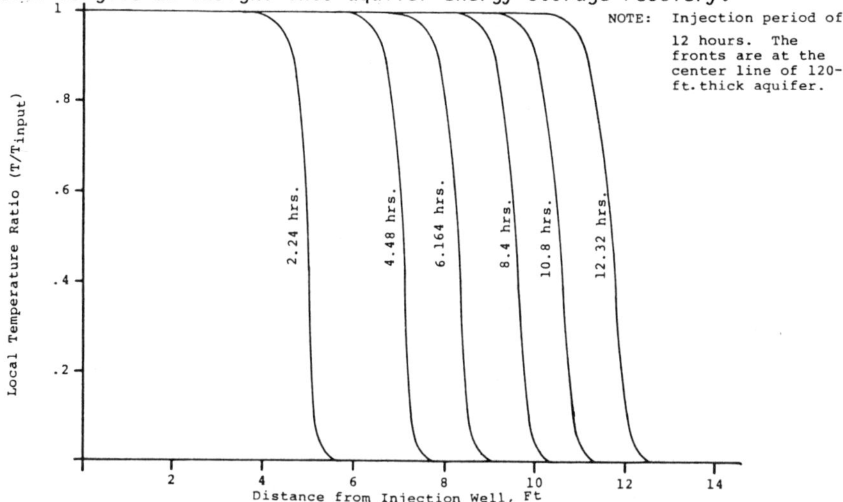

Fig.4.7. Ground water thermal fronts during injection into an unconditioned aquifer (based on a daily cycle).

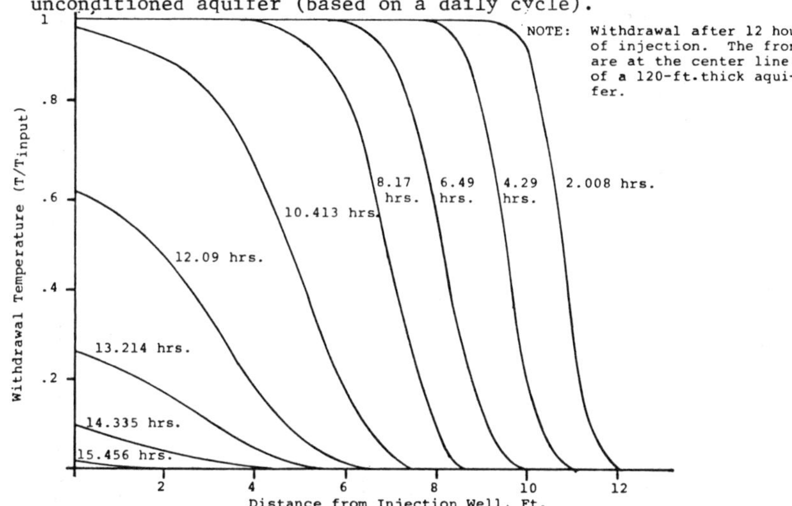

Fig. 4.8. Ground water thermal fronts during withdrawal from an unconditioned aquifer (based on a daily cycle).

58

Figures 4.7 and 4.8 show the centerline temperatures for injection and withdrawal for a 24-hour cycle in a 120 ft thick unconditioned aquifer. For a conditioned aquifer (six-month previous injection), the variation in recovery temperature is negligible from the input temperature along the centerline. This variation is not shown.

Figures 4.9 and 4.10 show the withdrawal temperature for an unconditioned and conditioned aquifer for an eight-day cycle. For the unconditioned aquifer no temperature decrease can be seen (for the scale utilized) until almost 75 percent of the water had been withdrawn. The temperature depreciation is rapid after this point. Based on a 75 percent temperature criteria, an 88 percent recovery results. For the conditioned aquifer, the temperature scale had to be changed to show any variation. The temperature decrease is only 0.5 percent after withdrawing the input energy.. This shows that an aquifer can be conditioned to achieve essentially complete temperature recovery.

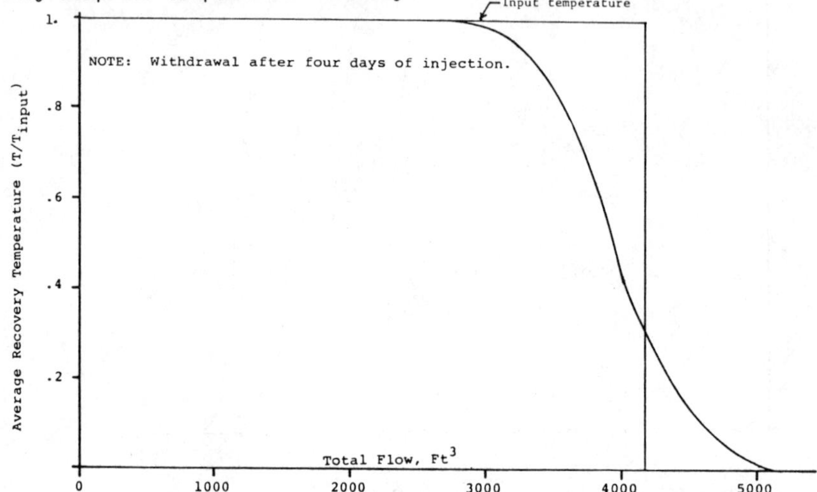

Fig. 4.9. Average withdrawal temperature versus total flow for an unconditioned aquifer (based on an eight day cycle).

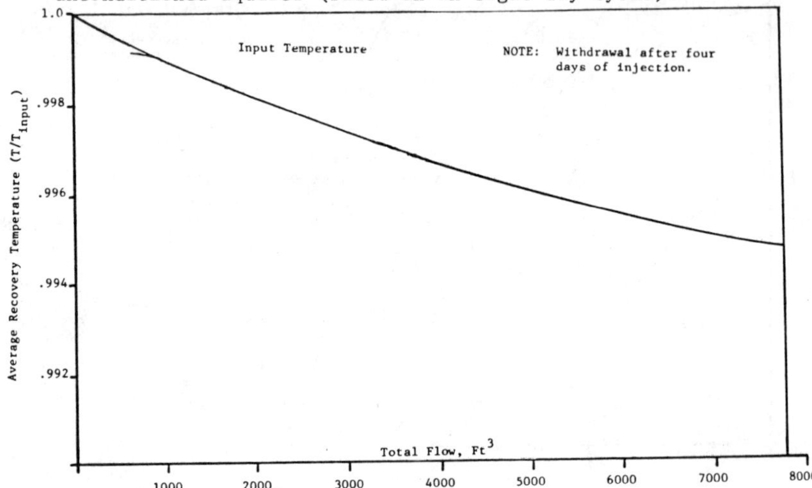

Fig. 4.10. Average withdrawal temperature versus total flow for a conditioned aquifer (based on an eight day cycle).

Figures 4.11 and 4.12 show the centerline temperature profiles for injection and withdrawal for an annual cycle in an unconditioned aquifer. The conduction effects resulting in residual energy in the aquifer are clearly demonstrated. Figure 4.13 gives the average recovery temperature and a curve of injected temperature versus quantity of water. The drop in temperature occurs relatively earlier than in the several-day cycle, but shows a gradual decrease. Figure 4.14 shows the centerline recovery temperature for a conditioned aquifer. On the temperature scale utilized, no degradation during recovery can be detected.

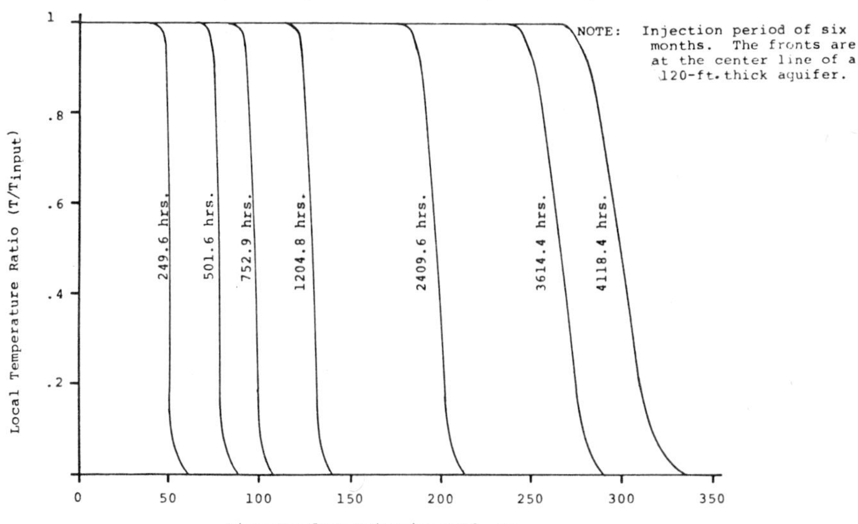

Fig. 4.11. Ground water thermal fronts during injection into an unconditioned aquifer (based on an annual cycle).

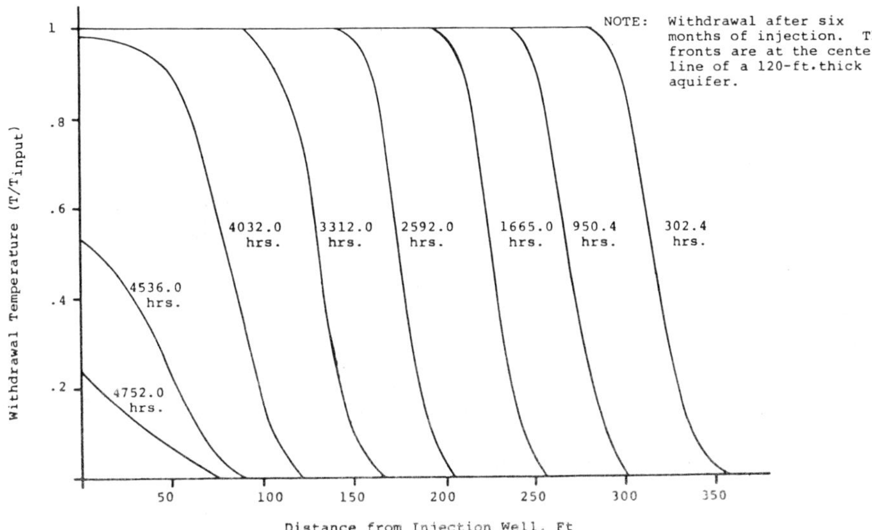

Fig. 4.12. Ground water thermal fronts during withdrawal from an unconditioned aquifer (based on an annual cycle).

60

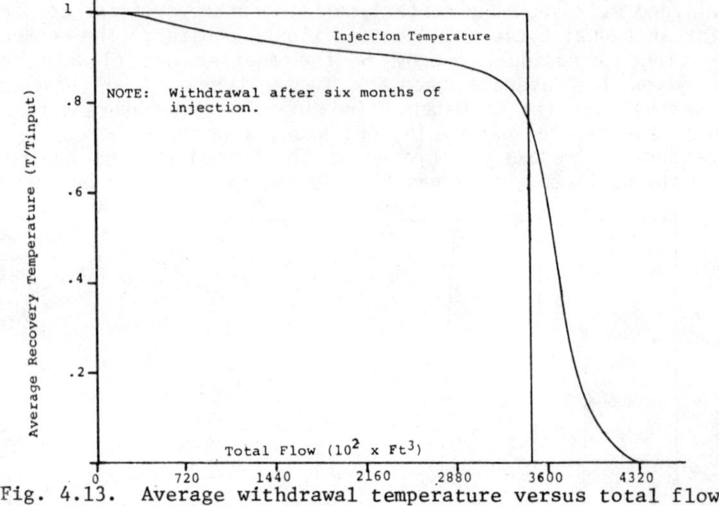

Fig. 4.13. Average withdrawal temperature versus total flow
for an unconditioned aquifer (based on an annual cycle).

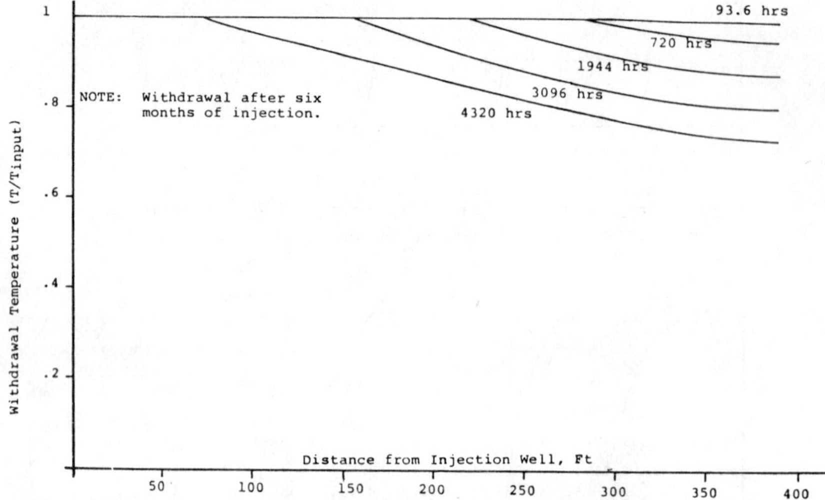

Fig. 4.14. Ground water thermal fronts during withdrawal from a
conditioned aquifer (based on an annual cycle).

The computer program utilized gives the temperatures in grid form both in the
aquifer and confining layers. Any injection-withdrawal time sequences can be
utilized. Injection temperatures are assigned and average withdrawal temperatures
are calculated. The output represents energy storage in an homogeneous aquifer.
Real aquifer performance calculations must be made with logical engineering
corrections.

Thermal storage computer programs are limited to use with large computers. The
calculations were made on a UNIVAC 1110 using up to 120,000 storage spaces. One
annual cycle has up to one hour's machine time. Minicomputer programs are not
available for the major thermal storage program. The results in the form of
efficiency terms can be utilized.

Example: Energy is to be stored in an aquifer 50 ft thick with a porosity of 0.3. Determine the storage volume and well spacing for a two-well system to store 10^9 Btu with a 50°F temperature difference.

Solution: The specific thermal energy storage is:

$$q = (\rho c_p \phi)_{water} + (\rho c_p(1-\phi))_{rock}$$

$$= (62.4 \text{ lb/ft}^3)(1 \text{ Btu/lb°F})(0.3) + (2.6 \cdot 62.4 \text{ lb/ft}^3)(0.22 \text{ Btu/16°F})(0.3)$$

$$= 43.7 \text{ Btu/ft}^{3}\text{°F}$$

The required volume is:

$$V = \frac{Q}{q\Delta T \eta_A} = \frac{10^9 \text{ Btu}}{(43.7 \text{ Btu/ft}^3\text{°F})(50°F)(0.68)}$$

$$= 673{,}000 \text{ ft}^3$$

The aquifer thermal energy recovery efficiency for 1 cycle for an aquifer 50 ft thick is utilized (Fig. 4.6).
For a 50 ft thick aquifer the horizontal area is:

$$A = \frac{V}{b} = \frac{673{,}000 \text{ ft}^3}{50 \text{ ft}} = 13{,}460 \text{ ft}^3$$

For a two well system (Eqn. 3-24):

$$A = 1.05 \text{ R}^2 \text{ or}$$

$$R = \sqrt{A/1.05} = \sqrt{13{,}460/1.05} = 113.2 \text{ ft}$$

This means the two wells must be spaced 113.2 ft apart. A spacing of 115 ft or 120 ft is reasonable. With a safety factor, a larger spacing can be used: however, excess spacing increases the pressure loss. Storage varies by the square of the distance between wells so a spacing of 150 ft gives 75 percent excess storage.

Natural Flow Effects

The natural flow in aquifers displaces the stored energy in the direction of the natural flow. Natural flow velocities are generally relatively low varying from a few feet per year up to a few feet per week. Most velocities are in the lower part of the above range. The effect on energy recovery can be losses which equal a fraction of a percent to 100 percent of the energy stored. Through proper placement of the wells the loss can be minimized. The movement of energy within the storage volume could possibly be considered environmental contamination.
Again returning to the thermal wave concept, the velocity which affects energy movement is the natural flow thermal velocity and not the natural flow water velocity. Thereby from Equation 4-9, the effective natural flow velocity is:

$$V_{Eff\ N.F.} = \frac{(c_p)_{water}\ \phi}{(\rho c_p)_{water}\ \phi + (\rho c_p)_{rock}(1-\phi)}\ V_{N.F.}$$

The following table gives some values of the effective natural flow velocity as a function of porosity and natural flow velocity. A specific heat of 0.22 Btu/lbm°F and a specific gravity of 2.6 is used for the rock. Standard conditions are used for the water.

		Natural Flow Velocity (ft/yr)				
Porosity		1.0	3.0	10.0	30.0	100.0
Porosity	$V_{Eff\ N.F.}/V_{N.F.}$	Effective Natural Flow Velocity (ft/yr)				
0.05	.084	0.084	0.25	0.84	2.5	8.4
0.10	.163	0.163	0.49	1.63	4.1	16.3
0.20	.304	0.309	0.91	3.04	9.1	30.4
0.30	.428	0.428	1.28	4.28	12.8	42.8
0.40	.538	0.538	1.61	5.38	16.1	53.8
0.50	.636	0.636	1.91	6.36	19.1	63.6

For example if the natural flow velocity is ten ft per year, the effective natural flow velocity is only 4.28 ft/yr for a porosity of 0.30. The movement of the thermal field is less than the movement of water field. The energy is stored in the rock and water, whereas the water fills only the voids.

A natural flow efficiency is defined as one minus the loss fraction or:

$$\eta_{NF} = 1 - Losses_{NF}$$

This allows an efficiency value to be used directly in the well placement calculations.

A linear approximation is utilized to estimate the energy losses. The effective natural flow velocity is multiplied by the cycle time to determine the storage displacement. The fraction of energy lost is equal to this distance divided by the distance across the storage volume in the same direction as the natural flow.

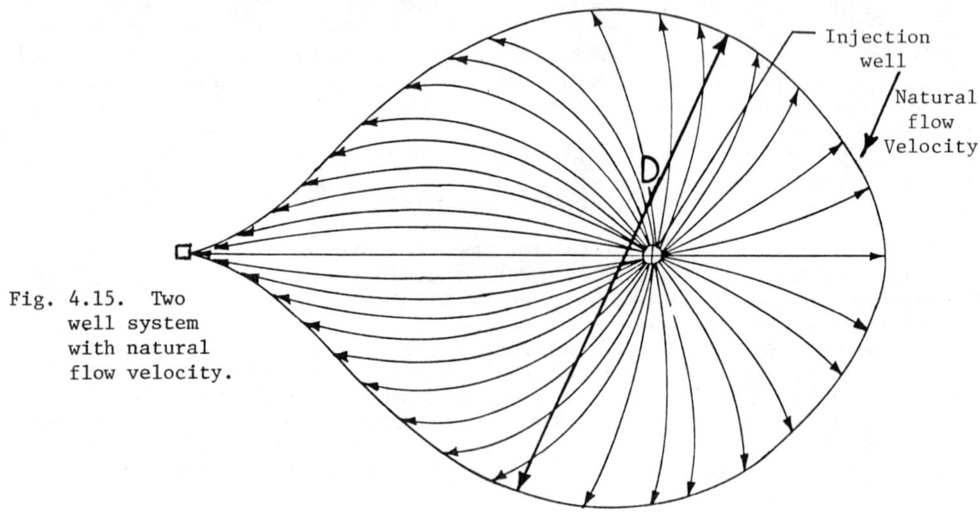

Fig. 4.15. Two well system with natural flow velocity.

This is best described by an illustration. Figure 4.15 shows the storage available for a two-well system. The natural flow velocity direction is designated at the upper right hand corner of the storage area. The wells are spaced 400 ft apart and the natural flow velocity is 10ft/yr with a porosity of 30 percent. The effective natural flow velocity from the previous table is 4.28 ft/year. The annual cycle displacement distance is 4.28 ft. The maximum distance "D" across the storage area in the same direction as the velocity is 420 ft. The distance is measured from Fig. 4.15, using the distance between wells as 400 ft. The fraction of energy lost is 4.28ft/420ft or approximately 1.1 percent. The effect of the natural flow is minor in this case.

For these calculations the sharp tip at the sink should be rounded off as noted in Fig. 4.15. If groups of paired wells are used as noted in Fig. 3.12 and 3.13, the energy loss to natural flow can be minimized by orienting the wells so the maximum length of the storage area is parallel to the natural flow. Losses due to natural flow normally will be a few percent at the most.

Thermal Water Conditioning

In a two-well system the water is thermally conditioned. If water at 200°F is required and the aquifer is at 60°F, the water must be heated from 60°F to 200°F in the first cycle. If the useful temperature range is 200°F to 130°F, only half the original energy input can be utilized. Half the energy is lost. In a two-well system the rejected water at 130°F is injected in the second well and stored. For the second cycle, neglecting aquifer losses, the water is heated from 130°F to 200°F. The water has been thermally conditioned.

Thermal water conditioning is not available with the single-well system. The water must always be heated from the source temperature to the required temperature. During the recovery stage the hot water (thermal pollution) is discarded. Thermal water conditioning is available for both hot water and cold water storage. The system works identically with chilled water.

General Design Sequence

To determine the aquifer energy storage system parameters, data on energy storage capacity, peak injection and withdrawal rates, and temperature variations are required. In most heating and cooling applications only the peak loads are normally calculated. In addition the aquifer parameters of thickness, physical size, porosity, and permeability are required. In general the aquifer parameters will vary throughout the aquifer. To compensate, logical engineering variations of the techniques established for a homogeneous aquifer can be applied.

A good aquifer is a prerequisite for a satisfactory aquifer thermal energy storage system. A general library survey will usually note the type and extent of aquifers in a region. Each state has a geological survey (by many names) and a branch of the United States Geological Survey. In many regions test well data are available. From this survey some areas will be rejected, some will be questionable, and some will have a very high probability for a usable aquifer thermal energy storage system. Since well production is very site-specific, specific site data are required. If a test well does not exist, one or more should be drilled.

Data from the test well are produced from well logs, geological samples from various depths, and a pumping test as described in Chapter V. Well logs will give the accurate location of the water-bearing material or aquifer. The sample log will identify the type of material and values for porosity and perhaps permeability. A pumping test of a few days duration at the required peak flow rates will verify a good well and suitable aquifer. The well diameter will be a function of required flow rate. A minimum of six-inch casing and sand screen is required for commercial-type water pumps. With proper packing this can perform as a ten-inch well. The drilled hole size is normally ten inches. This well size is normally sufficient for 250 gpm. For higher flow rates larger wells should be utilized. In most applications the test well can be converted to a production well when drilled at the proper location.

By drilling three test wells and measuring the head in each well from a common level, the natural flow direction can be determined. The natural flow will be in the direction of the greatest head gradient. If the permeability and porosity are known, the natural flow velocity can be determined. An example of a natural flow calculation is given in Chapter III.

In normal heating and cooling applications, the heating, cooling, and ventilating systems are analyzed for peak loads. These are the input for the system design. When annual energy storage is utilized the annual heating and/or cooling loads are also design requirements. The peak loads will be several times the average load. Designing a system with excess capacity results in increased capital cost and operating cost. Maintenance cost, many times given as a percentage of captial cost, will be higher and in some cases energy usage will be higher. A pump operates at peak efficiency near a single flow rate. Operating at one-fourth or one-third of the peak efficiency rate can cut efficiency in half. Peak efficiency flow rates are noted in the pump tables of Chapter VI.

Once the aquifer properties and thermal requirements are known, the storage system can be determined. The sequence is:

    a. Determine specific thermal capacity:

$$q = (\rho c_p)_{water} \phi + (\rho c_p)_{rock} (1-\phi)$$

    b. Determine aquifer efficiency η_A from Fig. 4.6:

    c. Determine volume from thermal storage load, $q\Delta T$ and: η_A

$$V = \frac{Q}{q\Delta T \eta_A}$$

A safety factor can be used at this point, however if η_A is for one cycle, a safety factor is built into the system for future years. The effect of a conditioned aquifer could also be included at this point; this would increase the efficiency.

    d. Determine horizontal area required from aquifer thickness:

$$A = \frac{V}{b}$$

    e. Determine well spacing required for one or more well pairs:

$$R = \sqrt{A/1.05} \quad (1 \text{ Well Pair})$$

See Equation 3-25 for more than one well pair:

    f. Determine effective natural flow velocity:

$$V_{Eff\ N.F.} = \frac{(\rho c_p)_{water} \phi}{(\rho c_p)_{water} \phi + (\rho c_p)_{rock} (1-\phi)} V_{N.F.}$$

g. Determine natural flow efficiency:

$$\eta_{N.F.} = 1 - (V_{Eff\ N.F.})(t)/D$$

h. Determine volume from:

$$V = \frac{Q}{q\Delta T\eta_A\eta_{N.F.}}$$

i. Repeat steps d and e (Volume change should be minor):

j. Determine aquifer head loss between wells:

$$\Delta h = \frac{Q\ \ln\ ((R/2)/r_w)}{\pi Kb}$$

If head loss Δh is too large, the number of well pairs must be increased. The pressure loss is proportional to the flow rate. Doubling the number of wells will reduce the pressure loss by almost a factor of two. More than one well pair may be required to supply maximum well flow rate capacity. The maximum flow rate is determined by peak loads.

Example Set of Calculations

To heat a group of buildings requires an annual load of 10^{10} Btu with a peak load of 10^7 Btu/hr. The useful energy range is 150°F to 100°F. Test wells have a maximum flow rate of 250 gpm per well. A 120-ft thick aquifer has a porosity of 0.20 and a permeability of 10^2 ft^3/day. Natural flow velocity is 10ft/yr. Design a reasonable aquifer thermal energy storage system.

a. $q = (\rho c_p)_{water}\phi + (\rho c_p)_{rock}(1-\phi)$

$= (62.4\ lb/ft^3)(1\ Btu/lb°F)(0.3) + (2.6 \times 62.4\ lb/ft^3)(0.22\ Btu/lb°F)(1-.3)$

$= 43.7\ Btu/ft^3°F$

b. $\eta_A \approx 0.75$ (Fig. 4.6, First Cycle)

c. $V = \dfrac{Q}{q\Delta T\eta_A}$

$= \dfrac{10^{10}\ Btu}{(43.7\ Btu/lb°F)(50°F)(0.75)}$

$= 6.10 \times 10^6 ft^3$

d. $A = \dfrac{V}{b} = \dfrac{6.10 \times 10^6 ft^3}{120\ ft}$

$= 50,800\ ft^2$

66

  e. At this point a determination is made to determine the number of wells required for peak load.

$$\dot{q} = \dot{m}c_p \Delta T$$

$$= 250 \text{ gal/min} \cdot 8.34 \text{ lb/gal} \cdot 60 \text{ min/hr} \cdot 1 \text{ Btu/lb}^\circ F \cdot 50^\circ F$$

$$= 6.25 \times 10^6 \text{Btu/hr}$$

$$Q_{Required} = 250 \text{ gal/min} \cdot \frac{10^7 \text{ Btu/hr}}{6.25 \times 10^6 \text{Btu/hr}} = 400 \text{ gal/min}$$

This flow rate is not sufficient to meet peak loads. To meet peak loads a four-well (two-well pairs) system is required. Using three well pairs increases reliability as the system could function with a well pair out of commission. Using the well configurations in Fig. 3.12 and the area coefficient of 1.1 (Fig. 3.13) with the distance between paired wells to distance between wells ratio of 2.0 gives:

$$A = \frac{50,800 \text{ ft}^2}{3 \text{ well pairs}} = CRD = 1.1 \frac{R^2}{2}$$

$$R = 175 \text{ ft}$$

  f.

$$V_{Eff\ N.F.} = \frac{(\rho c_p)_{water}\phi}{(\rho c_p)_{water}\phi + (\rho c_p)_{rock}(1-\phi)} V_{N.F.}$$

$$= \frac{(62.4 \text{ lb/ft}^3)(1 \text{ Btu/lb}^\circ F)(0.2)\ 10 \text{ ft/yr}}{(62.4 \text{lb/ft}^3)(1 \text{ Btu/lb}^\circ F)(0.2)+(2.6 \cdot 62.4 \text{ lb/ft}^3)(0.22 \text{Btu/lb}^\circ F)(1-0.2)}$$

$$= 3.04 \text{ ft/yr}$$

Maximum distance of storage area is perpendicular to the well pairs or approximately 500 ft. The efficiency from natural flow is:

$$\eta_{N.F.} = 1 - V_{Eff.N.F.}\ t/D$$

$$= 1 - (3.04 \text{ ft/yr})(1 \text{ yr/500 ft})$$

$$= .994$$

This will not change the area so additional calculations for area are not required.

  k. Maximum head loss through aquifer between wells is:

$$\Delta h = \frac{(Q/3 \text{ well pairs}) \ln (\frac{R}{2}/r_w)}{\pi K b}$$

$$= \frac{((400 \text{ gal/min})/3)(\frac{60.24 \text{ min}}{\text{day}})(\frac{1 \text{ ft}^3}{7.48 \text{ gal}}) \ln((\frac{175\text{ft}}{2})/0.25\text{ft})}{3.14 \ (10^2 \text{ft}^3/\text{day})(120 \text{ ft})}$$

$$= 3.99 \text{ ft}$$

This head is satisfactory. The final configuration is six wells with a distance of 175 ft between wells in a well pair and 87.5 ft between well pairs.

Summary

The techniques developed for calculating thermal energy storage in aquifers apply to well-defined ideal homogeneous aquifers. The thermal capacity of aquifers is almost independent of aquifer variables where there is a small variation as a function of porosity. These thermal energy estimates are valid for most aquifers. The pressure of large fractures in aquifers could change thermal capacity. The flow of water through the aquifer is very dependent on aquifer homogeneity. In most cases water flow parameters must be determined from test data. Once the flow is determined, the thermal energy storage system can be developed.

References

1. Baumeister, T., and Marks, L.S., _Standard Handbook for Mechanical Engineers_, McGraw Hill, 7th Edition, (1967).

2. Perry, J.H., _Chemical Engineers Handbook_, McGraw Hill, 4th Edition, 1963.

3. DeWiest, R.J.M., _Geohydrology_, Wiley, N.Y.

4. Tsang, E.F., Buscheck, T., Mangold, D., and Lippmann, H., "Mathematical Modeling of Thermal Energy Storage in Aquifers," in _Thermal Energy Storage in Aquifers Workshop_ , LBL-8431, UC-94a, CONF-7805140, May 1978.

5. Davison, R.R., Harris, W.B., and Reddell, D.L., "Texas A & M Cold Water Aquifer Project, "_Proceedings of Solar Energy Storage Options_, CONF-790328-P1, March, 1979.

6. Ansari, J.M., _High Temperature Thermal Energy Storage in Aquifers with a Solar Power Plant Application_, Ph.D. Dissertation, The University of Alabama, (1980).

7. Ebeling, L., Reddell, D.C. Tostengard, S., Harris, W.B.,and Davison, R.R, "The Effect of System Size on the Practicality of Aquifer Storage," _Proceedings of ISES Meeting_, (1979).

8. LeCroy, J.E., _Thermal Storage and Recovery in a Model Aquifer_, M.S. Thesis, The University of Alabama, (1980).

CHAPTER V

WELLS

Well Drilling Methods

The book of Genesis, in the 26th Chapter, 18th verse, states "Isaac digged again the wells of water, which they had digged in the days of Abraham..." The earliest records of drilled wells in China go back to the Chou dynasty (1122 to 256 B.C.) and in Mesopotamia, between the Tigris and Euphrates rivers, ancient civilization progressed commensurate with the quantity of water available from wells. Doubtlessly most wells were dug by hand, although it is recorded that the Chinese employed spring-pole power, the forerunner of the percussion method that has survived until modern times.

The well is the conduit between the aquifer and the surface. Methods of drilling and completing water wells are dictated by a number of factors such as depth, type of rock penetrated, aquifer characteristics, expected yield, and cost. These variables also relate to the diameter of the hole and the casing program. Geologic conditions place wells in two broad categories--first, wells that penetrate relatively "soft" soil and rock and are completed in unconsolidated water-bearing alluvium or sand and gravel above bedrock; and second, wells that penetrate and are completed in indurated rock such as sandstone, limestone, or dolomite. In some instances, metamorphic and igneous rocks containing factures or other voids may serve as suitable aquifers.

A well penetrating a water-bearing section comprised of consolidated rock generally will employ a short string of casing set through the unconsolidated soil and overburden with the hole left open below the casing shoe. If the aquifer consists of unconsolidated alluvium or sand and gravel, casing will be set on bottom through the aquifer and either perforated or slotted, or a screen may be attached to the lower end of the casing and set opposite the water-bearing section.

Although several methods of constructing water wells are in use, commonly the cable-tool procedure or the rotary method of drilling is employed. In instances where the geologic section is relatively "soft" and the aquifer moderately shallow, wells of large diameter are bored with rotary augers. The choice of equipment is governed by one or more of the factors noted above. Of these, cost is likely to be the primary consideration.

Cable-Tool Operations

The cable-tool method of drilling is often referred to as percussion or standard method and, as noted, has a long history of use in developing groundwater supplies. Percussion tools were employed as early as 600 B.C. by the Chinese for drilling brine wells to depths of a few hundred feet. Following the completion of the Drake oil well in Pennsylvania in 1859, cable-tool technology advanced rapidly and improvements were made in tools and power sources, but the basic concept remains unchanged since early times.

The cable-tool rig makes hole by lifting and dropping a string of tools suspended on a cable or drilling line. The tools consist of a drill bit on bottom attached to a heavy drill stem by a tool joint, followed by a set of drilling jars attached by a second tool joint to a swivel or rope socket which is, in turn, connected to the drill line (Fig. 5.1). The drill line consists of a left-lay steel cable, usually 5/8-inch or 7/8-inch in diameter, passing over a sheave in the crown block at the top of the mast downward to the walking beam which imparts the reciprocal motion to the line and drilling tools. From the walking beam the cable passes to the bull reel where it is wound and fed off as hole is made. The walking beam is connected by a pitman to a crank which is activated by the rig power system.

The cable is held taunt in the downward motion, but the lay of the line permits it to unwind and stretch slightly, thus imparting a turning action to the bit which causes the cutting face to strike the bottom of the hole at a different position at each stroke. This results in a circular hole and assists in maintaining the vertical attitude of the drilled hole.

The jars consist of a pair of linked bars with several inches of vertical play. The play in the jars permits an upward blow to assist in freeing the tools when they become stuck in the hole. Otherwise, the jars serve no purpose except that of assisting in delivering the full momentum of energy to the drill bit on the down stroke.

.Water in nominal quantities is introduced into the hole to mix with the rock particles cut by the bit, which forms a slurry that must be removed from the hole periodically. This is accomplished by use of a bailer or a sand pump. The bailer consists of a section of pipe, generally four inches to six inches in diameter, with a check valve at the bottom. The sand pump is similar but is fitted with a plunger that creates a partial vacuum on the up-stroke to suck the slurry into the pump body above the check valve. The bailer or sand pump is attached to a sand line that passes over a sheave on the crown block to a sand reel where it is wound.

In drilling unconsolidated formations, it is frequently necessary to install a string of casing to contain the walls of the hole. Casing is driven down the hole at intervals of some five to ten feet of hole drilled. Another function of the casing is to shut off excess water that may be encountered before total depth is reached. Excess water will reduce the ability of the tools to make hole.

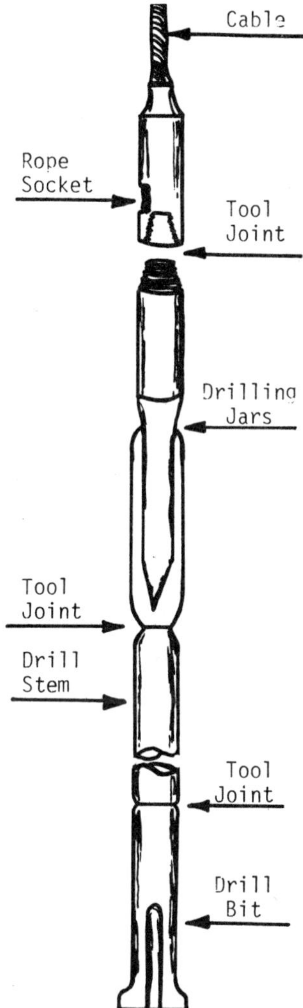

Fig. 5.1. Cable tool string.

Advantages of the cable-tool rig are several in number. Compared with rotary tools, the initial capital investment is lower as are daily operating costs. Water requirements are nominal, the rig is highly portable and suited to rugged terrane, excellent cutting samples of rock penetrated are obtained, and aquifers are easily indentified and evaluated. There is little risk of contamination of water-bearing zones. (Fig. 5.2).

Disadvantages of the cable-tool system include the relatively slow rate of penetration and the limitation of depth and hole diameter. Other disadvantages are the requirement for casing when drilling unconsolidated strata and the inability to control flow from formations penetrated. Formation evaluation by use of special logging techniques is limited to gamma and neutron devices where holes are cased.

Fig. 5.2. Trailer mounted cable tool rig.

Rotary Drilling Method

Rotary drilling is accomplished by use of a rotating drill string in conjunction with a pressurized circulating system employing fluids, either mud or air, to remove drill cuttings from the hole. Pressure is applied by conventional pumps or air compressors to force the drilling fluid down a hollow-stem drill shaft out through nozzles in the bit and back to the surface in the annulus between the drill string and the wall of the hole. The rotary method was developed in the United States in the 1800's and became commonplace after the drilling of the Lucas oil well with rotary tools at Spindletop near Beaumont, Texas, in 1901.

The basic rotary rig consists of a derrick or mast, a hoist, power plant, circulating pump or pumps, rotary table, drilling shaft, derrick equipment such as a swivel, traveling block, crown block, and wire lines, and the drill bit (Fig. 5.3). Weight is placed on the bit by use of one or more thick-walled drill collars as the lowermost part of the drill string or, in the case of rigs of shallow capacity, by a hydraulic pull-down mechanism. Two basic types of bits are employed--the roller type or the drag type. The roller bit consists of, commonly, three cones with cutting teeth of various designs on spindles and roller bearings set at an angle to the axis of the bit. This is the bit normally employed to drill hard formations and is referred to as a rock bit. A second-type bit is the drag bit used to cut softer formations. The fishtail bit consists of two short blades set at a slight angle toward the direction of rotation whereas the three-way design employs three cutting blades. There are several variations in the design of the drag bit that are related to its specific use.

Fig. 5.3. Portable water well rotary rig.

The bit is attached to the drill collars which are, in turn, attached to the drill pipe. Drill pipe in general use consists of seamless steel tubing 2 3/8 to 4 1/2 inches in outside diameter in lengths of 20 to 30 feet fitted at one end with a tool-joint pin and at the other with a tool-joint box. At the top of the drill string is the kelly, a square or hexagonal-shaped joint of pipe about ten feet longer than the individual joints of drill pipe employed. The kelly passes through drive bushings installed in the rotary table which imparts rotation to the entire drill string. As hole is drilled, the kelly slips downward through the drive bushings to maintain the bit on the bottom of the hole.

The kelly is attached to a swivel suspended by the drilling line from the crown block to the traveling block. A flexible, high pressure rotary hose is connected by a gooseneck to the swivel to transmit the drilling fluid to the drill string from the hydraulic pumping system. The dead-end of the drilling line is secured at the base of one leg of the derrick or mast and the other end passes over the sheaves of the crown block and through the sheaves of the traveling block and thence to the drum of the hoist which feeds the line to the drill string as drilling progresses. The line is wound on and fed off the hoist alternately in removing or returning the drill pipe from and to the hole in making a round trip to replace the bit or for other operations.

Rotary rigs of medium or low capacity may employ only a single power plant to operate the rotary, hoist, and hydraulic systems. Rigs with greater depth capacity usually employ one power plant for the hoist and rotary table operation and a second plant for the circulating system. Steam power was used in early days of rotary drilling, but was replaced by gasoline engines beginning in the 1920's and later by diesel engines. Gasoline engines are popular today for powering water-well rigs.

A direct hydraulic circulating system consists of a mud pump (or pumps) with a large-capacity suction line immersed in an earthen mud pit or in a large-capacity steel mud tank. The drilling mud passes from the pump through a steel mud line to the rotary hose, thence through the swivel to the drill pipe. The return flow passes from the well head through a shale shaker or screen which diverts the cuttings into a storage pit and the main flow to the opposite end of the mud pit or pits where it is picked up by the suction. Samples of cuttings for lithologic analysis are collected at the shale shaker (Fig. 5.4).

The capacity and power requirements of the pump are determined by the size of the drill pipe and size of the hole, the required volume of drilling mud expressed in gallons per minute (gpm), and the mud velocity needed to bring the rock cuttings to the surface. Mud volume requirements may range upward from a few hundred gpm to 1,000 gpm or more. Fluid velocities may be of the range of 100 to 200 feet per minute, depending on the size of the drill pipe, size of the hole, and pressure applied. Surface pressures ranging from 100 to 500 and more pounds per square inch may be required to maintain the needed volume and velocity of flow.

Drilling Fluids

Drilling fluids serve several purposes, primarily those of removing cuttings as they are produced by the bit, transporting the cuttings to the surface, and the cooling of the bit and lubrication of the bit bearings. Other functions include support of the wall of the hole to prevent caving, the prevention of fluid entry into the hole from porous and permeable formations encountered, the prevention of drilling fluid loss into highly permeable zones, the lubrication of mud pump pistons, liners, and rods, and the reduction of corrosion of the drill pipe. In the case where air is the circulating fluid, compressors replace mud pumps, and air transmits the cuttings to the surface and cools the bit. Large volumes of air at high velocities are required. In the drilling of deep wells where large quantities of water are encountered and where geologic formations are incompetent,

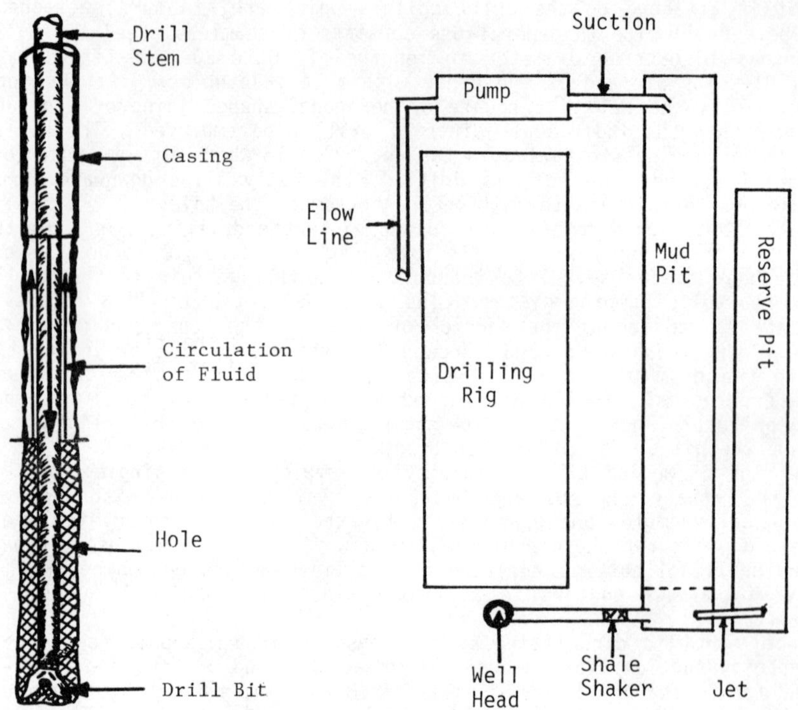

Fig. 5.4. Mud circulating system.

drilling with air is not feasibile. Circulating systems employing air are generally highly satisfactory in drilling relatively shallow water wells. One advantage of an air system is that it becomes immediately evident when water-bearing strata are encountered.

Water-base mud is the common circulating medium. Fresh water introduced at the surface mixes with clay particles derived from shales encountered as drilling progresses to form a satisfactory drilling mud. At shallow depths, before a natural clay mud is formed, it may be necessary to add commercial clay such as bentonite to the system to increase the viscosity of the drilling fluid to the point where cuttings are readily brought to the surface. In the drilling of shallow water wells the quality of the drilling mud is of little concern as long as noxious or polluting substances are not introduced into the aquifer. Deep wells, however, may require treatment to control both the chemical and physical properties of the mud requisite to the successful completion of the entire drilling operation.

Proper control of the quality of drilling mud is a science unto itself, and requires skilled technicians called "mud engineers". Water is the major constituent of most drilling fluids, and its general quality can be measured as a function of the hydrogen-ion concentration, or pH, which is the logarithm of the reciprocal of the hydrogen-ion concentration in moles per liter. A pH of seven, as measured on a scale of one to fourteen, is neutral, neither of high acidity or alkalinity. Acidic water below a pH of seven should be treated with soda ash (Na_2CO_3) to increase the pH to eight or nine before the introduction of mud additives. Soda ash is used to treat muds contaminated with gypsum and anhydrite; it removes the calcium ion from solution. Caustic soda (CaOH) is a strong alkali.

It is used in small amounts along with a thinner such as tannins, certain phosphates, or lignitic material to increase the pH of acidic water and prevent corrosion and the growth of micro-organisms. Thinners reduce the viscosity and gel development in muds without lowering the mud density. Water, of course, is the more common thinning agent used to reduce the viscosity of muds containing a high concentration of native clay.

The role of clay additives to drilling fluid is related to their colloidal behavior. Colloidal clay such as bentonite becomes hydrated with many times its own weight of water, forming a gel which prevents cuttings from settling to the bottom of the hole when drilling is interrupted. Colloids have good filtration properties, that is, they deposit a thin filter cake of low permeability on the walls of the hole allowing only small amount of water filtrate to invade permeable formations. The filter cake tends to seal the wall of the hole and prevent caving, thus permitting long strings of casing to be run in the hole without encountering obstructions formed by bridging. Also, the filter cake serves to prevent loss of mud into highly permeable zones that may result in "lost circulation" or "lost returns".

Density of drilling mud is expressed in terms of weight in pounds per cubic foot or pounds per gallon. One cubic foot has a volume of 7.48 U.S. gallons. One gallon of fresh water weights 8.34 pounds and, therefore, one cubic foot weights 62.4 pounds A column of fresh water exerts a pressure of 0.433 pounds per square inch (psi) per foot of height, or 43.3 psi per 100 feet of depth. This is pressure, or formation pressure, exerted by fluids at various depths within the Earth's crust. Thus a weight of 8.5 to 9.0 pounds per gallon of drilling mud would contain fluids within formations and prevent entry of such into the drill hole. This range of drilling-mud weight is considered adequate in drilling water wells to depths reaching 1,000 to 2,000 feet.

Higher pressures may be encountered in deep formations. In such cases it may be necessary to add a weight material to the mud column such as barite $BaSO_4$, which has a specific gravity of about 4.25 compared with that of water. Clays have a specific gravity of about 2.4 to 2.6, which serves to increase the weight of drilling mud to a range of 9.0 to 9.5 or more pounds per gallon, depending on the percentage of clay in the mud.

Viscosity is an important property of drilling mud; this is defined as a measure of internal resistance of a fluid to flow, and is an important factor in transporting drill cuttings and other solids to the surface. A balance must be maintained between viscosity, mud density, filtration properties, gel strength, and chemistry of the mud. Weight must be adequate to offset formation pressures encountered and to support the walls of the hole, yet not impair the efficiency of the mud pumps. Viscosity must be maintained at a level that will remove drill cuttings without impeding the rotation of the drill string, and the filter cake must be capable of sealing the walls of the hole, yet not build up a thickness that will cause undue restrictions in removal of the bit from the hole or "mud off" potential producing zones. Use of appropriate additives and chemicals will serve to control these factors and maintain the drilling mud in the best condition to serve the important role it plays in the successful drilling of wells.

Field testing of drilling fluids employs standard procedures developed by the American Petroleum Institute. Test equipment is simple to operate and provides satisfactory results, although it should be emphasized that testing in the field does not supplant more detailed procedures that properly find their place in the laboratory.

The Mud Balance offers a simple method for determining density of drilling mud. It is similar to an ordinary scale with a graduated arm resting on a knife-edge fulcrum, a cup to contain the mud attached to one end, and a rider that slides along the arm to indicate the weight of the mud in pounds per gallon or per cubic foot when the instrument is in balance. Viscosity is measured by use of a Marsh Funnel Viscosimeter, a funnel-shaped device with a precise orifice through which the flow of mud into a graduated cup is measured and reported in the number of seconds required for one quart of the sample to run out.

Filtration properties are measured in a low-pressure Standard Filter Press which employs a small cylinder of compressed nitrogen as its pressure source. A hardened filter paper is placed in the bottom cap of a cell, the cell is filled with drilling mud, the top cap secured, and a pressure of 100 psi applied for 30 minutes. The filtrate is measured in a graduated glass cylinder at an outlet in the bottom cap and reported in cubic centimeters. The filter cake thickness is measured and reported in thirty-seconds of an inch. Qualities such as texture, hardness, and flexibility of the filter cake may be recorded. Hydrogen-ion concentration of the filtrate is measured by use of strips of pH paper which change color after immersion in the filtrate. The color is matched with a color chart and the pH recorded to the nearest 0.5 level. Sand content of the drilling fluid may be measured by sieve analysis.

Other parameters of the drilling-mud filtrate including chemical analysis for determination of hardness, alkalinity, and chloride content can be measured by use of field kits, but in practice the determination of the qualities of the drilling mud as noted above will serve to regulate drilling fluid characteristics for the solution of particular well-drilling problems.

Other Common Drilling Systems

Inasmuch as the majority of wells developing groundwater supplies are drilled with either cable-tool or rotary rigs, only brief mention of other drilling systems and variations will be made. Boring with earth augers is a fairly common practice where aquifers are at shallow depths, the rocks penetrated are relatively soft, and holes of large diameter are desirable. In drilling with a rotary bucket, the drilled material is collected in a cylindrical bucket fitted with auger-type cutting blades on the bottom. The drill pipe is rotated by employing a kelly driven by a rotary table. A second boring method is done with a continuous spiral auger, generally four to six inches in diameter and in five-foot lengths, turned by two rotary drive-heads and forced down and retrieved by a hydraulic-feed mechanism. This method is limited to areas where the overlying formations consist of clay, and is not applicable where unconsolidated sands are encountered.

Rotary drilling employing air as the circulating fluid has been mentioned previously. The drilling rig is comparable to rigs using mud as the circulating medium, but air compressors are substituted for mud pumps. Cuttings removal is provided by the buoyancy of the air stream and requires a minimum air velocity to accomplish this purpose. Air pressures of some 50 psi delivering 300 to 400 cubic feet per minute are adequate for drilling to depths of 200 to 300 feet. A variation of the air-drilling technique is that of incorporation of a percussion device or downhole hammer into the system. A reciprocating engine powered by air delivers a sharp blow to the bit to fracture the rock formation. Cuttings are removed by air circulation.

Reverse-circulation rotary drilling may be employed where it is essential to obtain good sample specimens of the rock penetrated. This procedure has been in use for many years in core-drilling operations designed to obtain geologic structural data. The flow of the circulating system is reversed--drilling fluid is introduced into the annulus between the drill pipe and the hole and cuttings and the return flow of mud move upward inside the drill pipe into the mud pit.

Driven wells can be installed in soft formations where the aquifer is at shallow depths, usually some 30 to 50 feet. Well points consist of steel tubing, generally two to three inches in diameter, fitted with a screen on the lower section and a pointed driving attachment at the bottom, and are driven by either hand power or rig power into the aquifer. Suction lift or jet pumps may be installed in most instances, dictated by the diameter of the tubing and the hydrostatic head of the water. Use of this technique is limited.

Formation Evaluation

Formation evaluation is the technology dealing with the derivation, analysis interpretation, and evaluation of the physical properties and fluid content of rock strata encountered in the drilling of a hole in the Earth's outer crust in search of petroleum, natural gas, water, and other natural mineral substances. From a broad viewpoint, the term "well logging" may be applied to this activity, and may further be defined as the operation wherein data with respect to lithology, petrophysics, fluid content, and other parameters of the drilled rocks are recorded and plotted versus depth. Cuttings samples and cores are described, rate of penetration of the drill bit is recorded, and observations made as to probable physical properties such as porosity, permeability, and the fluid content of a reservoir or aquifer. Wire-line tools referred to as electrical logging devices are employed in most instances to provide further qualitative and quantitative analysis of formations penetrated in the drilling operation.

The chances of success in the drilling of a well in search of water or other substances are directly related to the ability to identify and evaluate potential productive zones. This task is usually placed in the hands of a well-site geologist. An experienced geologist is able to classify cuttings from rocks penetrated as to type, mineral constituents, grain size, texture, degree of cementation, and apparent physical properties, thereby affording a reliable opinion as to probable fluid content and the chances of an adequate rate of flow into the borehole.

Rocks are classified as to original state, whether of igneous, metamorphic, or sedimentary origin. Igneous rocks are those solidifying from a molten magma rising from the Earth's interior and, except for certain volcanic surface flows found mainly in the northwestern states, are devoid of porosity and hence unsuitable as aquifers.

Metamorphic rocks consists of rock types that have undergone modification by heat and pressure and, again, are not common as aquifers.

Probably 99 percent of the rock strata encountered in drilled wells are of sedimentary origin. These are classified as clastic, or fragmented rocks made up of particles of other rocks, minerals, and organisms. Common unconsolidated clastic sediments include alluvium, sand, gravel, and some silts, and clays. Consolidated clastic sedimentary rocks include siltstone, sandstone, clay, shale, and conglomerate. The preponderant constituent of sandstones is quartz grains (SiO_2). Non-clastic sedimentary rocks are those comprised of chemical precipitates. These are principally limestone ($CaCO_3$), chalk ($CaCo_3$), and dolomite [$CaMg(CO_3)_2$]. Other precipitates are the evaporites salt ($NaCl$), gypsum ($CaSO_4.2H_2O$), and anhydrite ($CaSO_4$); these, however, along with clays and shales, are not classified as aquifers.

Lithologic or sample logs generally are plotted at the well site, and show graphically the thickness of each lithologic type together with a brief description of color, grain size, texture, mineral composition, and the apparent presence or absence of porosity. Cuttings are normally examined under a low-power binocular microscope. Cuttings from cable-tool operations and rotary tools employing air as the circulating medium are generally of better quality than samples obtained in rotary drilling employing mud as the drilling fluid. An experienced geologist,however, can evaluate rotary cuttings satisfactorily. A record of the rate of penetration of the bit is usually kept and is plotted graphically as a drilling-time log showing time versus depth for intervals of ten or five feet, or one foot of hole penetrated. This is a very useful log as it can determine the depth to and thickness of zones of contrasting hardness or softness which relate respectively to lack or presence of porosity. The log will correlate with the deflections of the traces of certain electrical logs.

Core samples of rocks penetrated provide an excellent means of evaluating the properties of aquifers. Portions of the cores are subjected to laboratory examination to determine values of porosity, permeability, and fluid content.

Porosity is a measure of the void space in a rock and is expressed as a percentage of the volume of pores per bulk volume of rock. This provides a volume measurement of the storage capacity of an aquifer. Permeability is a measure of the capacity of a rock to permit flow of fluids through the interconnected pore spaces, and is expressed in darcy units or, in groundwater terminology, as the laboratory coefficient of permeability. The latter (K_s) is defined as a flow of 1 gal/day through a cross-sectional area of 1 ft^2 with a hydraulic gradient of 1 ft/ft. One darcy equals 18.2 K_s. These parameters have been discussed more fully in a prior chapter.

Formation evaluation employing electrical logging devices (or borehole geophysics) is the technique of lowering sensing devices into a well and recording at the surface a number of formation parameters as the tool passes the various strata. Logging by this method was developed by the petroleum industry and is now widely applied in the water-well industry. Many techniques are used in the petroleum world, but only those applicable to water-well evaluation will be noted.

An electrical current passes through rock formations on account of the conductivity of mineralized waters in the rock pores. The reciprocal, or resistance to current flow, is measured by the Conventional Resistivity (specific resistance) Method which measures the resistance between opposite faces of a unit cube of substance at a specific temperature. The meter is chosen as the unit of length, so the unit of resistivity is the ohm-meter^2/meter or, simply abbreviated: ohm-m.

In resistivity logging, a current is transmitted by means of a cable from the surface to an electrode arrangement in the hole. The drilling fluid in the hole conducts the current into the Earth and a meter at the surface records the potential difference between a surface electrode and the well electrode as it passes the various formations. If a particular formation contains fresh water, resistivity of the formation is a high value. Salt water is a good conductor, and resistivity of a bed containing salt water is low in magnitude.

A second parameter is usually measured simultaneously with the resistivity log by means of another electrode on the logging tool or sonde. This is called Spontaneous Potential (SP), which is a measure of the naturally-occurring potential difference between a surface electrode and the well electrode in the mud column as it passes different formations. The downhole ohmic potential in the mud column results from natural currents of electrochemical origin flowing around the intersection of permeable beds and impermeable beds. This device, then, serves to identify non-permeable beds, generally shales, and permeable beds, generally porous sandstones and limestones. The unit of measurement is the millivolt, and the trace of the recorded value is deflected to the right, or positive direction, opposite shales, and to the left or negative direction opposite permeable zones.

The relationship between the resistivity of the drilling mud and that of the formation water influences the nature of the SP curve. Low mud resistivity with respect to the formation resistivity will result in a wide deflection of the SP curve. In the case of fresh-water formations where the conductivity of the mud is greater than that of the formation water, the deflection of the SP curve generally will be reversed, or toward the right. This is important in the detection of fresh-water aquifers. Fresh-water mud may result in little or no deflection of the curve opposite fresh-water aquifers; the addition of a small amount of salt to the drilling mud will increase the magnitude of the reverse deflection and thus assist in the identification of fresh-water zones. The SP log is also used to determine the resistivity of formation water and therefore provides one method of determining ionic concentration which relates directly to water quality.

Radiation logging devices are also useful in the identification of water-bearing zones. These are of two basic types—those that measure the natural radioactivity of formations (Gamma Ray Log), and those that measure radiation reflected from or induced into formations from a neutron source contained in the sonde (Neutron Log). Gamma rays are spontaneous electromagnetic waves emitted by the nucleon of radioactive atoms such as Uranium ^{238}, Thorium ^{232}, and the Potassium isotope K^{40}. Shales contain a concentration of radioactive salts greater than that of sandstones, limestones, etc. Therefore, the Gamma Ray Log is essentially a shale log. The recording trace of the log deflects toward the right when radioactivity increases, and is therefore similar to and may be correlated with the SP log which also indicates the presence of shales. Energies of the SP log are expressed in units of "million electron-volts" (Mev).

Gamma Ray devices may be run in both open and cased holes filled with any fluid, including air. Thus it is suitable for determining interfaces of shales and other strata in open holes filled with either fresh-water or salty muds, and in empty holes, either cased or open.

The neutron logging method involves the placing of a source of fast neutrons close to a radiation detector on the sonde that is lowered in the borehole. Neutrons are emitted at high velocities and are slowed by absorption or "captured" by atoms of hydrogen, silicon, sodium, etc., which produces an emission of high-energy gamma rays. These so-called "capture" gamma rays are recorded by the radiation detector. The unit of measurement is micrograms of radium-equivalent per ton (ug Ra eq/Ton).

The hydrogen atom has a mass very near that of the neutron and is highly effective in the slowing and capture of the neutron. If the hydrogen concentration in a formation is large, the neutrons are captured within a short distance from the source. Where the hydrogen concentration is small, the neutrons travel some distance from the source. Accordingly, the detector counting rate decreases opposite formations with high hydrogen content, and increases with decreasing hydrogen concentration. Hydrogen concentrations are relatively high in the pores of both water-bearing and hydrocarbon-bearing formations. If the formation is not clay or shale, the counting rate is directly related to the amount of water, oil, or gas contained in the pores, and thus is a measure of porosity.

The neutron log is calibrated in units of standard counts per second, and is usually run on the downhole tool along with the gamma-ray device. Deflections of the log which correspond to variations of porosity can be differentiated from those related to shale content by comparison with the gamma-ray curve and the SP curve. In the evaluation of formations encountered in the drilling of water wells, a combination of the resistivity and SP devices and the gamma ray-neutron devices generally will provide positive identification of aquifers. Figure 5.5 is a diagrammatic comparison of typical responses of logging devices to the characteristics of representative formations. Although portable equipment is available for use by the well driller, it may be prudent to employ the service of commercial logging companies whose personnel are trained in the interpretation of data provided by the various logging tools. (Fig. 5.5).

There are several logging devices that have not been discussed above. These include logging tools applicable in the water-well industry but those which may not provide sufficient additional information to justify cost. Micrologging devices identify porous zones; caliper logs record the diameter of the hole; sonic logs measure sonic-wave travel time in rock formations that provides a measure of porosity; acoustic devices reveal the quality of cement bonding between the casing and the hole; temperature logs will locate the depth to cement surrounding the casing; and more exotic radiation devices provide additional data. Commercial logging companies will recommend use of such tools in instances where the information derived may be of significant value in formation evaluation.

Casing

Casing installed in wells drilled for water or other fluids serves three purposes: (1) to prevent collapse of the borehole; (2) to exclude pollutants or other extraneous substances, either of surface or subsurface origin, from entering

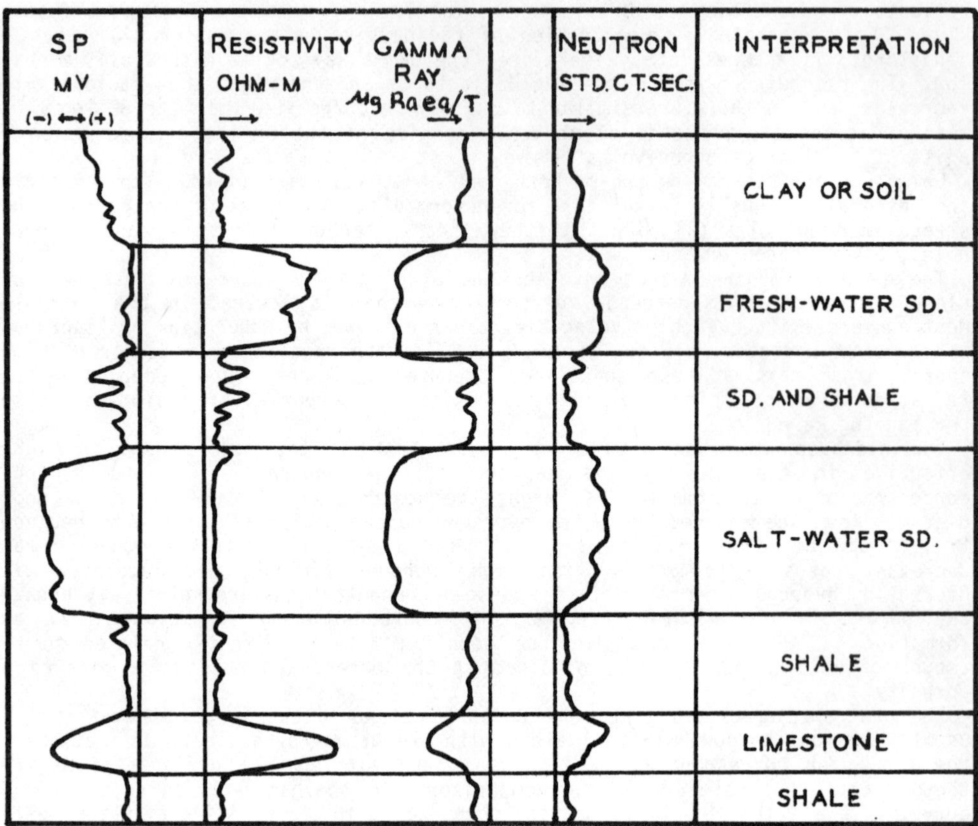

Fig. 5.5. Diagrammatic comparison of the response of various logging devices to the nature of typical rock formations.

the well; and (3) to provide a conduit to the surface for produced fluids via tubing and the downhole pumping equipment. Casing is of several types and designs, ranging upward from one to 1 1/2 inches in diameter for driven wells to as much as 36 inches for industrial and irrigation wells.

Steel pipe suitable for both shallow and deep wells is manufactured to specifications of several organizations, including the American Petroleum Institute (API). API standard 5L covers seamless or welded threaded and coupled line pipe suitable for most water-well installations. Other specifications used in water-well construction are designated ASTM A-120 and ASTM A-53; the former is made by the seamless or welded process, and the latter by either seamless, furnace butt-weld, or the electric-resistance weld processes. Table 5.1 is a listing of common line pipe nominal sizes showing actual outside diameter (OD), inside diameter (ID), wall thickness, and weight per foot threaded and coupled.

TABLE 5.1. Specifications for Line Pipe

Nominal Size (inches)	OD (inches)	ID (inches)	Wall (inches)	Weight/Foot (pounds)
1 1/2	1.900	1.610	0.145	2.75
2	2.375	2.067	0.154	3.75
3	3.500	3.068	0.216	7.70
4	4.500	4.026	0.258	11.00
6	6.625	6.065	0.280	19.45
8	8.625	7.981	0.322	29.35
10	10.750	10.192	0.279	32.75
12	12.750	12.090	0.330	45.45

Casing commonly in use in the petroleum industry is applicable in the water-well industry but is generally designed for greater setting depth and consequently is of higher tensile strength and collapse resistance. Table 5.2 gives selected examples of API seamless casing of various steel grades.

TABLE 5.2. API Seamless Casing

Nominal Size (inches)	Grade	ID (inches)	Wall (inches)	Weight/Foot (pounds)
4 1/2	H-40	4.090	0.205	9.50
4 1/2	J-55	4.000	0.250	11.60
5 1/2	J-55	4.892	0.304	17.00
6 5/8	J-55	5.424	0.288	18.00
6 5/8	N-80	5.921	0.352	24.00
7 5/8	H-40	7.025	0.300	24.00
7 5/8	J-55	6.969	0.328	26.40
7 5/8	N-80	6.875	0.375	29.70
10 3/4	H-40	10.050	0.350	40.50
13 3/8	H-40	12.715	0.330	48.00

Round-thread seamless API non-upset tubing is suitable for the pumping system. Typical examples are shown in the following table.

TABLE 5.3. API Non-Upset Tubing

Nominal Size (inches)	OD (inches)	Wall (inches)	Weight/Foot (pounds)
1 1/2	1.900	0.145	2.75
2	2.375	0.167	4.00
3	3.500	0.216	7.70
3 1/2	4.000	0.226	9.50
4	4.500	0.271	12.60

Thread specifications for various types of line pipe, casing, and tubing differ materially. API casing threads are rounded, eight threads per inch, with a taper of 3/4-inch per foot in diameter. API tubing differs in that tubing below four inches in diameter has ten round threads per inch. Water-well casing is likely to have sharp threads and a different taper. Particular attention must be given to thread dimensions of various types of pipe, as an attempt to join differing threads may likely result in a casing failure.

Use of polyvinyl-chloride plastic well casing (PVC) has become widespread in the water-well industry in recent years where the setting depth does not exceed 200 feet. Rubber-modified polystyrene and epoxy plastic pipe are being used for setting depths to 300 feet. Plastic casing is available in sizes ranging from 1 1/2 to 6 inches in diameter. Advantages of plastic casing are that it is corrosion-resistant, of light weight, and generally less costly than steel casing. Couplings are not threaded.

Grouting of Casing

Grouting or cementing of well casing involves the placing of cement between the casing and the borehole, or between two or more strings of casing where more than one string is used. Grout is commonly a mixture of Portland cement and water to form a slurry, generally in the proportion of five to six gallons of water per 94-pound sack of cement. This will produce a slurry with a weight of fifteen to sixteen pounds per gallon; weight can be reduced to some 14 pounds per gallon by the addition of 1 1/2 to 2 1/2 percent bentonite to the mixture, which is preferred for most applications. Bentonite helps hold cement in suspension, reduces shrinkage, and increases fluidity of the slurry.

In cementing casing, a continuous operation is required so that the initial set of the cement does not occur before the job is completed. In cementing casing in wells of the depth range of a few hundred to several or more thousand feet, it is expedient to employ the services of a commercial cementing company. Surface equipment is designed to be capable of mixing approximately 50 sacks of cement per minute, or larger volumes where two or more mixing units are operated in parallel. Cement is fed to the mixers either in sacks or from specially designed bulk hoppers.

The first step in cementing is to calculate the volume of slurry required to fill the annulus between the casing and the borehole to the desired height--either to the surface in the case of shallow wells or to 150, 200, or possibly several hundred feet depending on the particular circumstance. In wells no greater than 1,000 to 2,000 feet in depth, possibly 150 to 300 feet of cement behind the casing should suffice.

After casing has been lowered to bottom, the cementing pumps are connected and circulation between the casing and the borehole is established. It is advisable to pump a few barrels of water ahead of the slurry to counteract chemicals in the mud that might affect the physical properties of the cement. Generally a float shoe containing a ball-type check valve is attached to the bottom of the casing to prevent reverse flow after the cement has been pumped out the bottom of the casing and into the annulus. A float collar can be placed in the casing string one joint above bottom if it is desirable to ensure that cement fills the annulus near the bottom of the hole. These devices can be easily drilled out.

A flexible bottom-hole plug is inserted in the casing ahead of the cement slurry to displace the water or drilling mud and to separate the fluid from the cement. The slurry is introduced above the plug and the calculated volume pumped into the casing. The slurry is then followed by a second or top plug. The top plug, and the slurry, are then displaced by the introduction of water or drilling mud. When the top plug hits the float shoe, the pump stalls and circulation stops abruptly. Casing may be left open at the surface if the float shoe or float collar operates properly and there is no return flow. If there is a malfunction of the float, the weight of the cement slurry behind the pipe will displace the water or drilling mud in the casing and result in a flow back into the casing. This can be avoided by closing the casing at the surface.

A waiting period of from 24 to 48 hours or as much as 72 hours is required for the cement to set properly; this is governed by the type of cement employed. A period of at least 48 hours is advisable before beginning any operation inside the casing. After the appropriate waiting period, cement and cementing plugs can be drilled out where required. If drilling mud is used to pump the slurry to bottom, this can be displaced with fresh water, and completion operations may then be initiated.

Well Screens

Well screens are frequently required in the completion of water wells in unconsolidated aquifers such as sand and gravel. In the case of indurated aquifers comprised of firm sandstone or carbonate rocks, screens generally are not required. Completion may employ the open-hole technique where the aquifer is uncased, or slotted or perforated sections may be placed in the casing string opposite the aquifer to prevent caving of the borehole.

Well screens of various designs are available. These include machine-perforated or slotted screens (Fig. 5.6a) which may be modified by an outside wrap of wire gauge (Fig. 5.6b), or shutter screens (Fig. 5.6c), vertical louvered screens (Fig. 5.6d), and, probably the most popular type--the vertical-rodded, wire-wrapped screen (Fig. 5.6e). The latter consists of a cage of vertical rods outside a continuous-slot body wrapped by a spiral covering of cold-drawn wire. The wire is welded to the rods at intervals to form a strong, one-piece unit. These screens may be used in wells drilled by any method.

Screens are available in a number of diameters, the more popular being those of four-inch and six-inch diameter, and are usually manufactured in ten-foot lengths which can be joined where greater lengths are required. The four-inch nominal diameter screen has an OD of 3 3/4 inches and the six-inch has an OD of 5 5/8 inches which permits the screens to pass inside or telescope through four-inch and six-inch casing, respectively. One advantage of use of the telescope screen is that it can be retrieved and used again.

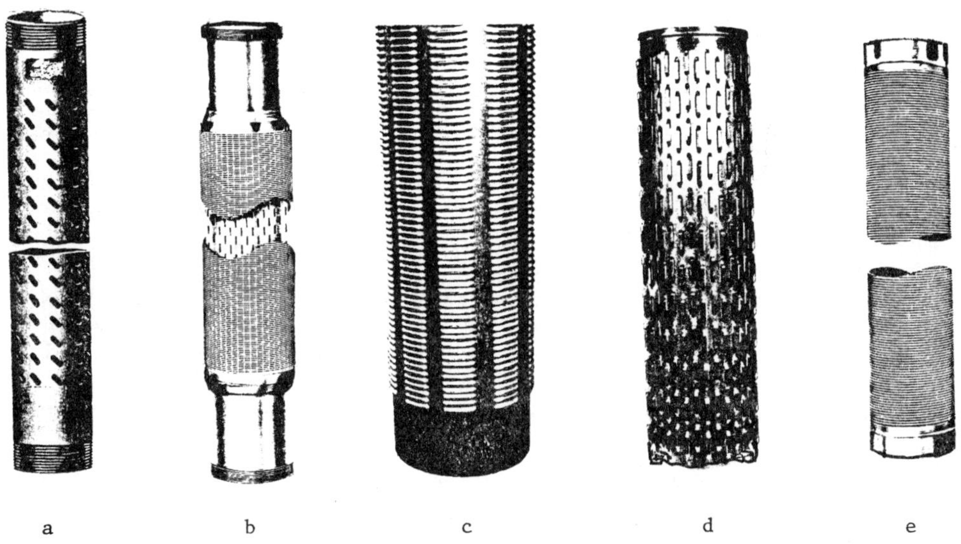

a   b   c   d   e

Fig. 5.6. Well screens.
(Courtesy of American Water Well Association)

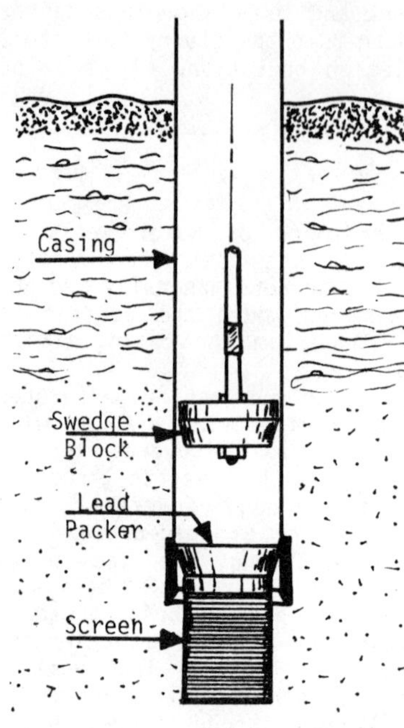

Fig. 5.7. Swedge block and lead packer.

The pull-back method of installing screens is particularly applicable in wells drilled with cable tools. The essentials of this method include lowering the casing to the bottom of the hole followed by lowering the screen to the bottom of the casing, then pulling back or lifting the casing to expose the screen to the water-bearing zone. The top of the screen is fitted with a lead packer which must be sealed after it is landed on bottom by use of a swedge block (Fig. 5.7). A second packer consists of a neoprene sealing ring at the top of the screen which is self-sealing and requires no expanding operation.

Although the telescope screen may be used in rotary-drilled holes, it is a common practice to attach the screen permanently to the bottom of the casing and set the casing and screen simultaneously. A wash-down fitting is installed at the bottom of the screen to permit removal of bottom-hole cuttings and heavy mud by circulating water through the casing and out the annulus. A bail-down procedure may be used to remove sand and cuttings from below the screen and permit the screen to settle to bottom as the cuttings are displaced, or a small-diameter wash pipe can be installed temporarily in the screen in conjunction with a back-pressure valve to circulate bottom-hole material out of the borehole and upward through the casing-borehole annulus. The back-pressure valve serves to prevent return flow into the screen and casing.

Gravel-Packed Wells

In some instances, an artificial gravel pack is placed between the screen and the borehole to provide stabilization of the aquifer. This may be accomplished by setting a casing string of large diameter on bottom, lowering the permanent casing and screen inside the larger casing, and pumping the selected gravel down the annular space. After filling with a few feet of gravel, the outer casing is pulled upward in steps until the level of gravel is well above the top of the screen. The outer casing is then removed. The permanent casing may be lowered to bottom inside the outer string, the gravel pack installed, and the screen telescoped to bottom if the telescope procedure is used.

Uniformly-graded gravel of a diameter three to five times that of the aquifer particles is used. This permits use of a screen with slots sized to retain only the gravel pack, and thus permits the installation of screens with much larger openings in aquifers composed of fine-grained material. Gravel-pack material graded in diameters of 1/16-inch to 1/8-inch generally is used, although some gravels of larger diameter are selected. A mixture of heterogeneous sizes should be avoided, as the grains may segregate as to size falling through the water or mud fluid with the result that the coarser grains fall more rapidly and reach the bottom before arrival of the finer material.

Gravel is sometimes dumped into the annular space between the casing and the hole in the case of shallow wells, but the chance is great that a separation of sizes will occur as the gravel settles through the water or mud fluid. This may result in bridging of the hole well above bottom, leaving a void space opposite the screen. This problem can be overcome by reverse circulation--pumping the gravel-laden fluid down the annulus where the gravel drops out on bottom and the fluid returns to the surface through the casing.

Well Costs

No infallable yardstick can be applied to the estimated cost of drilling and completing a particular well as a result of the many variables and contingencies involved. It is therefore necessary to resort to generalities and averages in arriving at estimates of cost considering the variables of well diameter, depth, completion procedures, equipment requirements, projected yield, geologic environment, and compliance with requirements of regulatory agencies. Well costs are a sum of providing labor, supplies, and equipment including the following: (1) moving in, (2) setting up, (3) drilling the hole, (4) expense of formation evaluation including electrical logging services, (5) casing, liners, screens, and fittings, (6) cementing, (7) certain consumables such as water and mud additives, (8) gravel packing, (9) testing operations, (10) installation of pumping systems and controls, and (11) moving out.

Variables in well costs, excluding pumping systems, differ primarily with respect to drilling method employed, type of geologic formations penetrated, diameter of hole, and aquifer characteristics. The following table provides average estimated cost ranges adjusted to 1979 levels for completed wells (excluding pumping systems) in several categories. Drilled wells employing nominal six-inch casing are considered capable of delivering an adequate volume of water for community-integrated heat pump systems.

TABLE 5.4. Estimated 1979 Cost of Completed Wells

Casing Diameter Inches	Aquifer Type	Drilling Method	Depth Feet	Cost Range Dollars
4	Sand and Gravel	Cable Tool	200	1,650-1,925
6	Sand and Gravel	Cable Tool	200	2,640-3,080
6	Sand and Gravel	Rotary	200	2,280-2,730
6	Sandstone and Limestone	Rotary	200	1,700-2,100
6	Sand and Gravel	Rotary	400	4,000-4,600
6	Sandstone and Limestone	Rotary	400	3,250-3,875
24	Sand and Gravel	Auger	50	950-1,200

It should be cautioned that the above cost figures are "ball park" estimates subject to many variables in a highly volatile market. The estimates must be adjusted to fit conditions in each specific case in various geographic regions, and in accord with the requirements of the well owner. An adjustment to the estimates may be made for the rate of inflation, but it follows that estimates or bids should be obtained from reputable contractors before preparing a budget for the drilling of a well.

Legal Aspects

In the drilling of wells for water or other substances, most states exercise regulatory control of the operation in one form or another. Many states issue drilling permits and require reports upon the completion of the well. In the case of wells drilled for water, there is considerable variation in the requirements of state agencies. Most all regulations have three functions, first a public health function to assure that wells are so constructed as to provide a safe drinking water supply, another designed to prevent pollution of groundwater through entry of polluted or mineralized water into aquifers, and a "consumer protection" function to ensure completion of an efficient well.

Control of the withdrawal and use of groundwater is practiced by many states. Most water-short western states operate under a prior-appropriation doctrine wherein water is allocated on a scale of need and beneficial use. Arizona, California and Utah are examples of states in which the withdrawal and use of groundwater is allocated on the basis of beneficial use. Priority is given to use for domestic and municipal supplies, followed by uses for irrigation, mining, recreation, etc. The water-rich states have generally adopted the riparian-rights doctrine under which little constraint is placed on the withdrawal and use of surface waters and, for all practical purposes, groundwater is considered personal property of the surface owner.

In Alabama, the use of water has not been subject to significant legislation and there is no control over water withdrawal. The Alabama Water Well Standards Board regulates completion practices, and requires that wells drilled for the public be drilled by a licensed driller. A permit is required for a water well, which is more or less automatically granted to the well driller. Florida law is administered by Water Management Districts which issue water-use permits based on availability and maximum reasonable-beneficial use. Georgia requires a permit for the withdrawal of more than 100,000 gallon of water per day. If there is sufficient evidence that groundwater is not consumptively used, a permit will be granted without a public hearing.

The State of Mississippi is the only state east of the Mississippi River that has adopted the prior-appropriation system for water rights. Permits for use of groundwater in problem areas are required. Legislation in Iowa states that a permit is required for using in excess of 5,000 gallons of water per day from any source of supply, but a municipal water system is excluded from the requirement.

Use of aquifers for thermal storage is a recent practice. Inasmuch as water withdrawn is returned to the aquifer with essentially no loss, this is not a consumptive use and, therefore, it is envisioned that no problem should be encountered in obtaining a permit for this particular use. Building and other local codes may apply to the drilling of water wells. In each case in every state, local and state authorities should be consulted to determine requirements and restrictions on the use of groundwater. The drilling contractor affords a source of reliable infomation with respect to all requirements.

Most state laws provide that protective measures be taken to prevent pollution of the water source. A final step in proper well completion is the sterilization procedure which is designed to destroy disease-producing organisms that may have been introduced into the well during construction operations via drilling fluids, equipment, materials, or by surface drainage into the well. A chlorine solution is an effective agent for this purpose. Calcium or sodium hypochlorite may be used as a source of the chlorine, or commercially prepared solutions are available for this purpose.

References

1. Brantly, J.E., Rotary Drilling Handbook, Palmer Publications, (1952).

2. Campbell, M.D., and Lehr, J.H., Water Well Technology, McGraw-Hill Book Company, (1977).

3. Campbell, M.D, and Lehr, J.H., Rural Water Systems Planning and Engineering Guide, Commission on Rural Water, Washington, D.C., (1973).

4. Johnson Division, UOP Inc., Ground Water and Wells, (1975).

5. Manual of Water Well Construction Practices, Environmental Protection Agency, EPA 570/9-75-001, (1976).

6. A Manual of Laws, Reguations, and Institutions for Control of Ground Water Pollution, EPA 440/9-76-006, (1976).

7. Well Drilling Operations, Department of the Army and the Air Force, TM 5-297/AFM85-23, (1965).

8. Well Drilling Manual, National Water Well Association, Worthington, Ohio.

9. Schaetzle, W.J., and Brett, C.E., Heat Pump Centered Integrated Community Energy Systems, NTIS No. ANL/CNSV/TM25, (1979).

10. Grubbs, D.M., and Cohen, H., "Legislature Framework for Water Resources Management in Alabama," Alabama Development Office, ALA-NRC-X996-WRD-73, (1973).

11. Legal and Administrative Systems for Water Allocation and Management, Proc. of a Southeast Regional Conference, Virginia Water Resources Research Center, Blacksburg, Va., (1978).

CHAPTER VI

PUMPING SYSTEMS

The pumping system transfers water from aquifer to the surface and in many cases directly to the place of application. The primary system components are the pump, controls and, in some designs, a pressure tank. The controls can be designed to regulate the flow rate directly through variable speeds and/or by controlling the number of pumps in operation. A second option uses a pressure tank. When the pressure tank reaches a preset pressure, the system shuts off. When the pressure decreases to a preset pressure the system starts up.

Pump Performance

Performance is determined from the hydraulic output of the pump divided by the work input. The hydraulic output of the pump is the total change in the flow work or pressure head (P/γ), the velocity head $(V^2/2g)$, and the potential head Z times the mass flow rate. The pump head is:

$$\Delta h = (\frac{P}{\gamma} + \frac{V^2}{2g} + Z)_{Discharge} - (\frac{P}{\gamma} + \frac{V^2}{2g} + Z)_{Suction} \tag{6-1}$$

The hydraulic power output of the pump is the mass flow times the pump head or:

$$P = \dot{m}\Delta h \tag{6-2}$$

The equation provides a _change_ in pump head, not the pump head. Using the proper conversion factors the pump output in horsepower is:

$$P(HP) = \frac{\dot{m}\Delta h}{33,000} = \frac{Q\Delta h}{3,960} \tag{6-3}$$

where

$\dot{m}$ is in lb/min
h is in ft , and
Q is in gpm

The pump efficiency is defined as:

$$\eta_p = \frac{Hydraulic\ Power}{Input\ Power} \tag{6-4}$$

The efficiency of the pump depends on flow rate, head, and specific speed. The specific speed for centrifugal pumps is:

$$n_s = \frac{n\sqrt{Q}}{\Delta h^{3/4}} \qquad (6-5)$$

where n is pump speed (rpm) and n_s is specific pump speed, Q is in gpm and Δh in feet. Typical pump parameter variations with flow rate are given in Fig. 6.1 for a centrifugal pump or axial flow pump. The peak performance is relatively flat over a range of flow rate. Pumps should be choosen so they operate near their peak efficiency. A computer-controlled system can dictate the number of operating pumps to achieve maximum efficiency. Pump tables that follow give the flow rate at which pump efficiency is maximum.

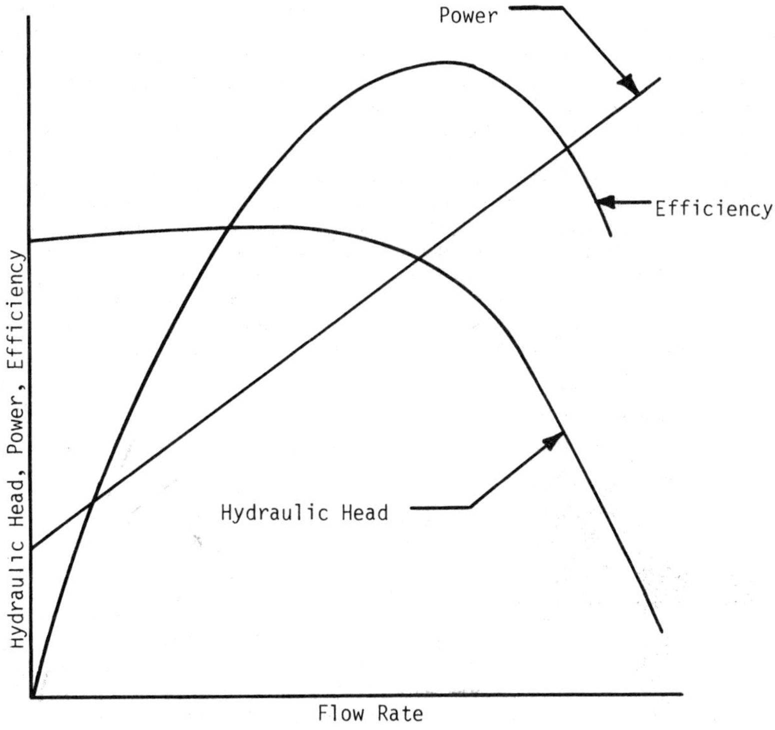

Fig. 6.1. Typical pump performance.

The peaks of the performance curves for above average pumps with Q gpm are:

$$n_p = 1 - \frac{0.8}{Q^{0.25}} \qquad (6-6)$$

As flow rate increases, pump efficiency increases. Pumps for aquifer thermal energy storage are expected to fall in the 50 to 90 percent efficiency range for flow rates of 20 gpm to 1000 gpm per well. The pump power required is the ideal power (Eqn. 6-3) divided by pump efficiency and motor efficiency or:

$$P_{actual} = \frac{Q\Delta h}{3960\ \eta_p \eta_m} \qquad\qquad (6-7)$$

The head loss includes pipe losses, aquifer losses, equipment losses, and head from water table to maximum height.

Pumps are classified in several ways--(1) positive displacement pumps which deliver the same volume of water within their operating range regardless of head, and (2) variable displacement pumps which deliver water in quantities varying inversely with respect to the water head. Within these two categories pumps are further classified as shallow-well and deep-well pumps, although this classification is inconsistent inasmuch as depth is a relative concept.

Most systems using aquifer thermal energy are expected to use a submersible, centrifugal-type pump. These pumps are available in a range of flow rates with corresponding pressure ranges. Due to simplicity and low maintenance requirements, economics are favorable for use of centrifugal pumps. In most household installations, a shallow-well pump is located at the surface and takes water by suction lift. Suction lift works by developing a partial vacuum in the intake tubing which causes atmospheric pressure to force water into the tubing and to the surface. This is possible only when the constant water head is less than 34 feet from the surface. Normal atmospheric pressure of 14.7 pounds per square inch (at sea level) is the equivalent of pressure exerted by a 34-foot column of water. Friction loss must be overcome, so that wells wherein the water head upon drawdown remains in the maximum range of 25 to 30 feet below surface can be pumped by suction lift. Vapor pressure, a function of temperature of the water, may be a factor in loss of pump efficiency; this is not significant at normal groundwater temperatures. In most aquifer storage systems, the depth will be above the 34-ft limit. To prevent the possibility of cavitation, pumps should be located well below the water level.

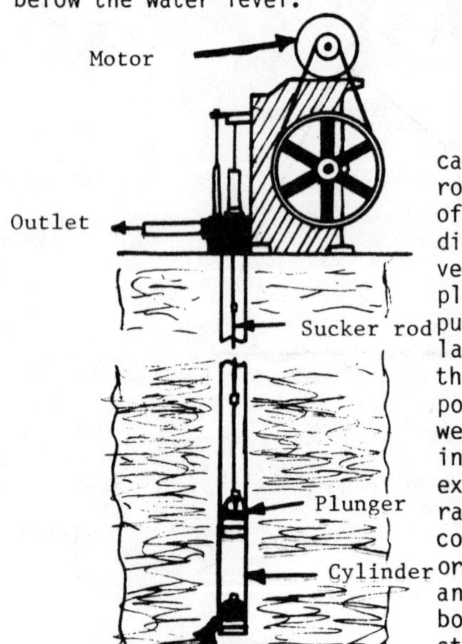

Fig. 6.2. Shallow well positive displacement pump.

Positive Displacement Pumps

Positive displacement pumps are divided into two categories, the piston-type pump and the rotary-type pump. Positive displacement pumps are of two basic types, (1) piston-type horizontal displacement surface installations and (2), vertical displacement submerged pumps fitted with plungers and check valves. Positive displacement pumps were used extensively in the past, but have largely been replaced by centrifugal systems in the water-well industry. An example of the positive-displacement pump still in use is in wells pumped with windmills. In the petroleum industry this type of pump is still used extensively to pump oil from wells at depths ranging up to a few thousand feet. The pump consists of a cylindrical body in which a piston or plunger attached to a plunger rod moves upward and downward in reciprocating motion. At the bottom of the pump is a check valve called a standing valve and a second valve known as the traveling valve is located within and is part of the plunger system. On the upstroke the standing valve opens to permit fluid entry into the pump while the traveling valve closes to lift fluid above it.   On the down stroke the standing valve closes and the valve in the plunger opens to permit fluid entry above it. The pump, attached to tubing and a string of pumping rods attached to the plunger rod, are lowered simultaneously inside the well casing. A second arrangement is commonly used; an insert pump is attached to the rod string and lowered inside the tubing to bottom where the pump is seated in a friction-type seating nipple. Figure 6.2 gives an example of a piston-type pump.

These pumps have flow capacities of up to a few thousand gallons per minute. Efficiency can be over 80 percent even at low flow rates (200 gpm) for high pressure heads (over 300 ft). The performance is relatively stable over a wide range of flow but is low for small pressure heads. Efficiency for a 50-ft head is below 50 percent at 200 gpm and around 60 percent for 2000 gpm.

The rotary type displacement pump is normally a screw or gear-type pump. These pumps have capacities up to 3500 gpm, and up to 1000 psi in pressure range. The efficiency is lower than that of centrifugal pumps; however, flow rate can be directly controlled.

Centrifugal Pumps

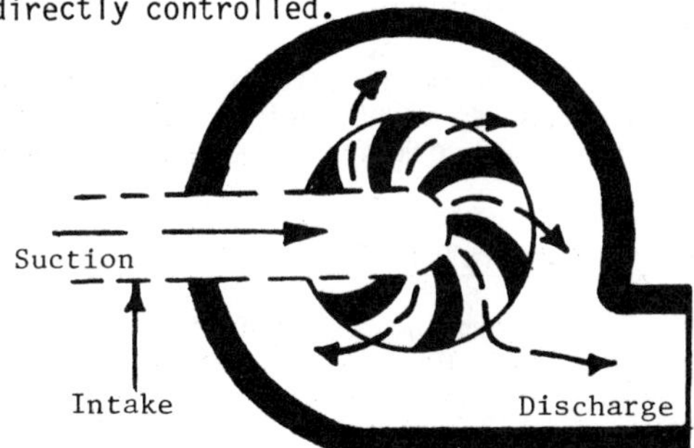

Fig. 6.3. Centrifugal pump.

Variable-displacement pumps in common use are generally of two types--the centrifugal and the jet type.   A centrifugal pump consists of a rotating impeller fitted in a case with intake and discharge openings. Figure 6.3 shows the basic workings of a centrifugal pump. The design of the impeller is such that its rotation forces water out the discharge and creates a vacuum drawing water into the impeller case through the intake opening. An advantage of this type pump is that an intake suction is created. In many cases this eliminates the need for priming. However, most larger pumps are primed.   Impeller design varies from those providing pure centrifugal pressure increases to those resulting in axial flow propeller type pressure increases. A mixed flow-type impeller imparting both centrifugal and axial flow compression is also used. The specific speed is around 500 rpm for pure centrifugal compression and reaches over 10,000 rpm for the pure propeller type compression. Figure 6.4 gives illustrations of impeller types and specific speed. The high specific speed models usually have smaller diameters, cost less, but have more wear.   The simplicity makes the centrifugal pump reliable and minimizes maintenance. Centrifugal pumps are available as single-stage pumps and multistage pumps. Figure 6.5 shows two configurations of multistage pumps and a pump mounted in a well. The pump capacity of each stage is independent of that of other stages. For example a pump head change from five ft to 105 ft is exactly the same as a pump head change from 205 ft to 305 ft.   Table 6.1 includes centrifugal pumps with four to eleven stages. Centrifugal pumps have high performance with efficiencies available within 90 percent of values derived from Eqn. 6-7.   The variations of power, capacity, and pump head with speed for centrifugal pumps are:

$$\frac{P_2}{P_1} = \left(\frac{n_2}{n_1}\right)^3 \qquad \frac{h_2}{h_1} = \left(\frac{n_2}{n_1}\right)^2 \qquad \frac{Q_2}{Q_1} = \frac{n_2}{n_1} \qquad (6\text{-}8)$$

The centrifugal pump used extensively in the water-well industry is the submersible type. The submersible pump is a centrifugal pump encased in a tubular body generally four inches to six inches in diameter closely coupled with an electric motor and with electric cables paralleling the flow pipe.   Water-tight

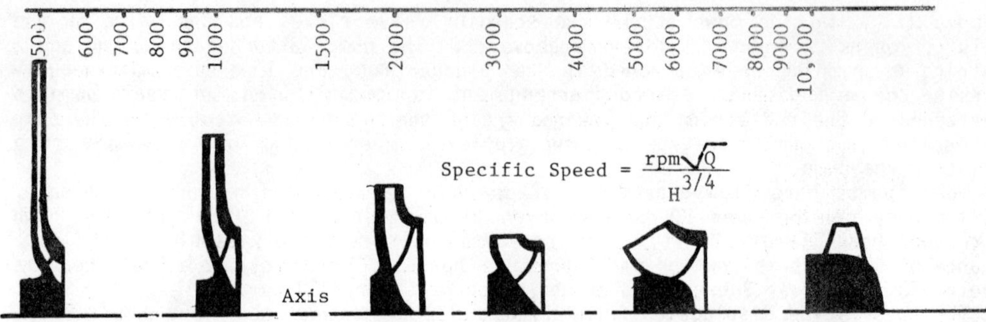

Fig. 6.4. Various impeller configurations.

seals permit use of the pumps to depths of 500 feet or more. The following table indicates performance and cost figures adjusted to 1979 prices of various combinations of power and capacity of a typical-make commercial pump of 6-inch diameter.

TABLE 6.1. Performance and Prices of Submersible Pumps 1979

Pump Cost	Number (1) of Stages	Delivery Pressure, psi	Pumping depth, feet						
			50	75	100	125	150	175	200
			Gallons/minute at indicated delivery pressure						
5.0 horsepower motor ($1,108) Controls ($215)									
$522	4	30	117	106	91*	72			
	4	60	75						
$678	8	30	90	85	81	77	71	65*	57
	8	60	78	75	66*	58	49	35	
7.5 horespower motor ($1,384) Controls ($215)									
$522	4	30	163	147	126*	108	80		
$659	6	60	113	105	95*	84*	70	52	
10 horsepower motor ($1,704) Controls ($275)									
$522	4	30	199	180	158*	129	90		
	4	60	141*	106					
$590	5	30	180	169	157	144	130*	114*	91
	5	60	150	136	120*	97	66		
$794	8	30	135	129	124	119	113	107	101
	8	60	122	116	110	104	97	89*	82
15 horsepower motor ($1,771) Controls ($369)									
$589	6	30	227	219	210	201	189	177	161
	6	60	205	194	183	167	147*	123	93
$722	8	30	198	191	183	177	169	161	154
	8	60	180	172	164	157	148	139	130
20 horsepower motor ($2,635) Controls ($478)									
$722	8	30	237	231	226	220	213	206	198
	8	60	223	217	210	202	194	184	174
$910	11	30	205	200	195	190	185	180	175
	11	60	192	187	182	177	172	167	161

(1) Stages refer to the number of vanes or impellers
* Flow rate at peak efficiency. Missing values are at higher pressures.
   All motors and controls for 3-phase, 230 volt service.

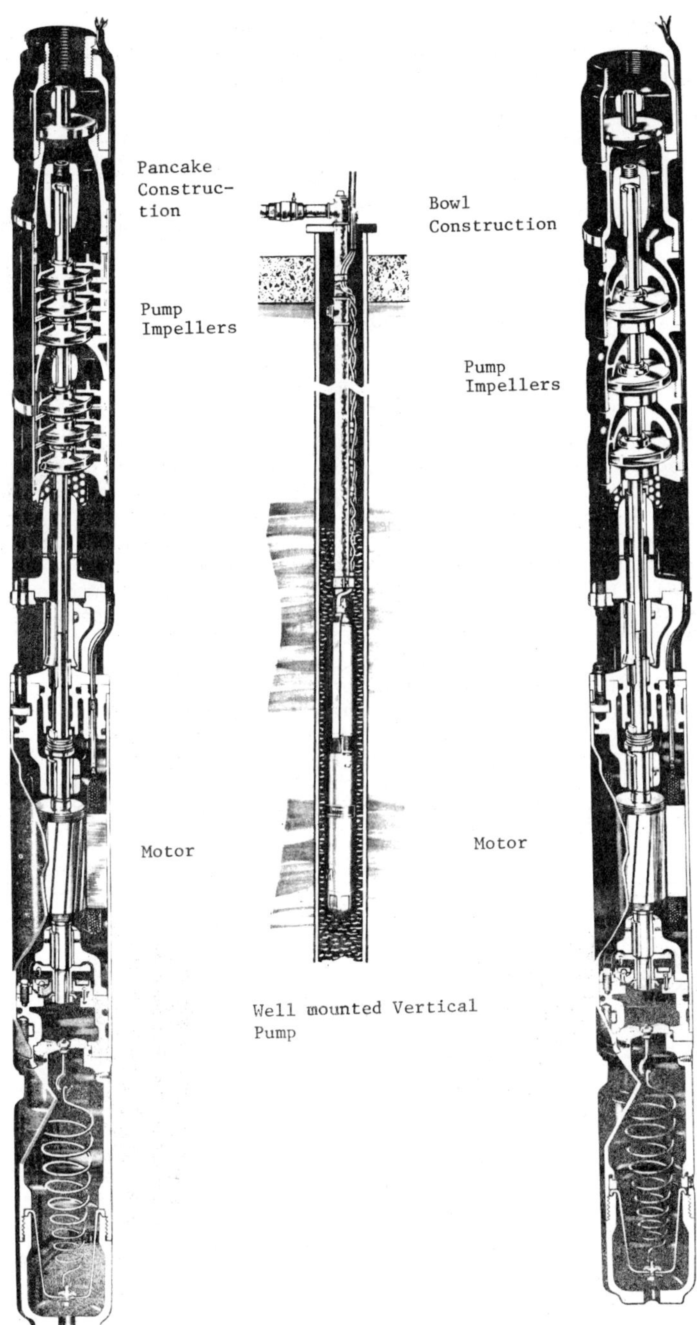

Fig. 6.5. Cross-section of vertical turbine pumps and a well mounted pump.

In domestic installations where large capacities are not required, submersible pumps for wells four inches or larger in diameter are available at moderate cost. Selected examples are shown in the following table.

TABLE 6.2. Selected Low-capacity Submersible Pumps

Cost (1979)	Motor HP	Stages	Delivery Pressure	Pumping depth, feet					
				60	80	100	140	220	300
				Gallons/minute at indicated depth					
$415	1 1/2	11	40-60	29	27	26	22		
$485	2	13	40-60				25	18	
$575	3	19	40-60					27	22

Jet Pumps

The jet pump is also a centrifugal pump but combines the use of an ejector or a Venturi-type jet and a nozzle. The pump is stationed at the surface and the jet mechanism is attached to the lower end of the intake pipe. A second string of pipe, of smaller diameter than that of the intake pipe called a pressure pipe, is run into the casing parallel with the intake pipe. The pressure pipe carries part of the flow from the pump discharge at the surface downward to the nozzle, and the increased velocity at the nozzle draws water into the venturi assembly. The combined flow of the circulating water and water increment from the aquifer is carried to the surface where part of the flow goes into storage and the volume required for the pressure line is recirculated.

Jet pumps by design are inefficient, but the several inherent advantages make the jet system highly adaptable to many installations. All moving parts are accessible at the surface, and the moving parts may be located some distance from the well head. The system is of simple design with relatively low initial and maintenance costs. Jet pumps are available in small diameters that permit installation in casing down to two-inch inside diameter. A schematic of a jet pump system is shown in Fig. 6.6.

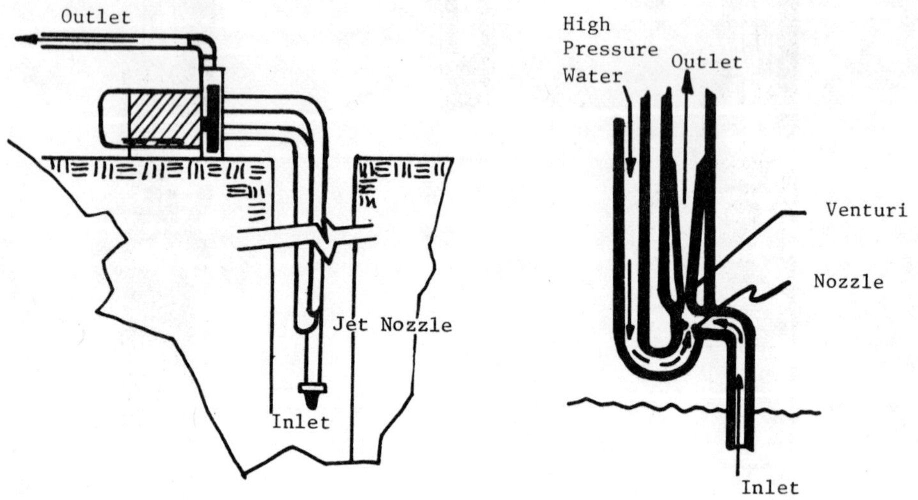

a. Jet pumping system   b. Jet ejector

Fig. 6.6. Jet-type pump system showing blow-up of jet nozzle.

Following are specifications for selected low-capacity jet pumps. Capacities may, of course, be increased with the selection of motors of greater power and system components of larger dimensions. Volume will be increased at lower delivery pressures.

TABLE 6.3. Selected Low-capacity Jet Pumps

Cost (1979)	Motor HP	Stages	Delivery Pressure	Pumping depth, feet					
				40	60	80	100	140	200
				Gallons/minute at indicated depth					
$210	3/4	2	40-60	8	6	4			
$220	1	2	40-60	11	11	9	6	4	3
$270	1 1/2	3	40-60	12	12	11	10	5	5

Economics

Pumping system economics include capital cost which includes first cost and depreciation, and indirect costs which include overhead and operating costs. The capital costs include pump, motor, controls, piping, foundations, freight, etc. Their value is measured by life expectancy and annual interest rate. Overhead includes engineering and inspection costs. Operating costs include energy, lubrication, and maintenance. Other important factors are pump availability, availability of maintenance service and parts availability.

The selection of a pumping system must be made in consideration of all variables as to depth, capacity, cost, etc. The logical source of information is the pump supplier or the drilling contractor.

Summary

The three handbooks listed in References give general information on pump capability and performance. Actual pump design is not pertinent to the material contained herein. In conclusion, it is expected that most systems will use submersible centrifugal pumps.

References

1. Karassik, I.J., Krutzel, W.C., Fraser, W.H., and Massina, J.P., Pump Handbook, McGraw Hill, (1976).

2. Baumeister, T., and Marks, L.S., Standard Handbook for Mechanical Engineers, McGraw Hill, (1967).

3. Fink, D.J., and Carroll, J.M., Standard Handbook for Electrical Engineers, McGraw Hill, (1968).

CHAPTER VII

COMMUNITY AQUIFER SYSTEMS

Systems using thermal energy storage in aquifers for residential or commercial heating and cooling may require the development of a public utility to manage transportation of the water with its thermal energy content. Physically the water distribution system will not be unlike the current utilities that distribute potable water and collect the waste water discharge. Such systems are most generally publicly owned, but a few private systems do exist, mostly in new residential or commercial communities. A similar ownership structure might be developed for community based thermal energy storage in aquifer systems.

This chapter considers the technical aspects of a community water distribution system. Economics and technical aspects normally require a community aquifer system for heating and cooling systems. The distribution system will be designed in a modular form. These modules can be readily expanded and interconnected to increase overall system reliability. Potential management and ownership structures for the community based water distribution system are also considered.

Water Distribution System

The water distribution system transfers the water between the aquifer and the individual energy users. The water distribution system is a piping network that must be both economical and compatible with the users' energy needs. To minimize flow rates in the distribution lines a modular system is a possibility. In this system a group of buildings are matched with a pair or pairs of wells as shown in Fig. 7.1. A community can consist of groups of modules as shown in Fig. 7.2. Matching modules give increased reliability as the failure of a pumping system in one module will be replaced by pumping in other wells. The need for emergency service is minimized. Using a module concept, the water distribution system can be started in any populated area and expanded to cover the entire community. The system is applicable to individual homes, multiple-family residences, commercial complexes, major buildings, etc.

In new construction areas, land use and building codes can be developed to require primary water distribution systems and wells. As there is additional population expansion in the area, new wells can be added to increase capacity. An additional set of wells will not only provide the increased requirements for one module, but also can be used to provide the increased capacity requirements for a number of modules. This type growth results from the specification of pipelines with growth capacity in the primary water distribution system.

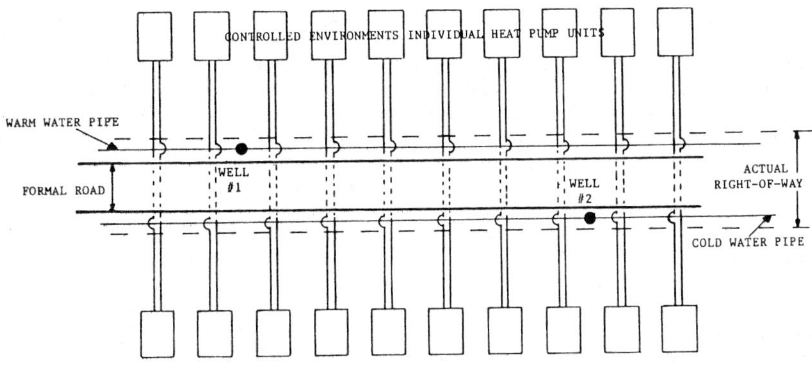

Fig. 7.1. Standard heat pump system module in heat pump integrated community energy system.

In downtown urban areas with major buildings, growth is expected to progress more quickly than in residential areas. Since retrofit is shown to be feasible, economic pressure will encourage the growth. Because modules will include fewer buildings, administrative decisions are minimized. Modules will be sized to provide sufficient water (well capacity) for a building or buildings. Many major buildings will be one module with groups of wells or could have more than one module with independent heating and cooling systems. Integrating with the other modules allows aquifer control and increases reliability. Connected modules need not necessarily be adjacent.

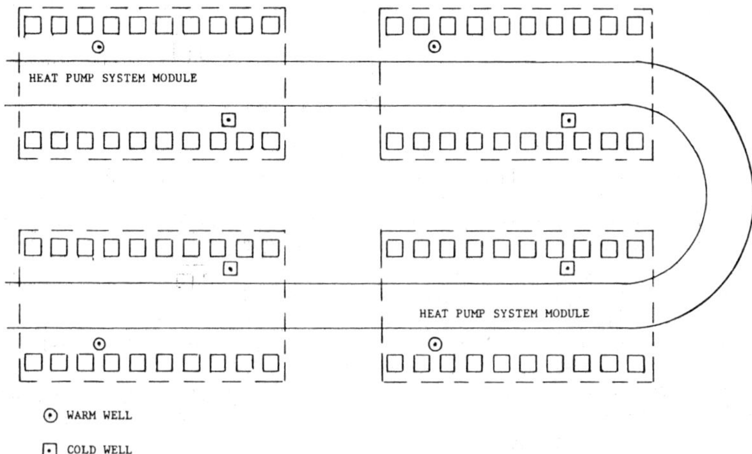

Fig. 7.2. Groups of heat pump system modules in heat pump integrated community energy system.

Piping Configurations

A number of ·different pipe configurations are available for the water distribution system. Figure 7.3 illustrates systems using one to four pipes. The systems duplicate those in major buildings for heating and cooling. The pipes are insulated in some cases and uninsulated in other cases, depending on the application. Water lines carrying $80^\circ F$ water ($27^\circ C$) or less are not normally insulated. Pipelines for higher temperature are normally insulated.

The first system (Fig. 7.3(a)) uses a one-pipe configuration between two community wells. During winter months water is withdrawn from the warm well and injected into the cold well. The direction of flow is reversed for summer months. A pump is used to circulate water through the heat pump for either heating or cooling. Performance is sacrificed in this system for the sake of economics. The first home along the pipeline from either well gets maximum performance.

To improve performance a two-pipe distribution system (Fig. 7.3(b)) can be utilized. One pipe will always have relatively warm water for heating and one pipe relatively cool water for cooling. During heating the water is cooled and rejected to the cool line and cool well. For cooling the water is heated and rejected to the warm line and warm well. If buildings are simultaneously being heated and cooled, some water will simply be transferred between buildings. Some heat will be rejected by one building and used by another building, without passing through the aquifer thermal energy storage system. The system minimizes well pumping by thermal diversity.

The two-pipe system can also be used for district heating or for a community solar heating system. One pipe is attached to the warm well and solar panel outlet and one pipe is attached to the cold well and solar panel inlet. The heat pump is replaced by a heat exchanger.

A three-pipe system (Fig. 7.3(c)) has a hot pipeline, a cold pipeline, and a return pipeline. This system could be used with a heat pump as shown or with direct heating and cooling.

The four-pipe system (Fig. 7.3(d)) separates the hot and cold systems. It can be used with a heat pump or direct cooling and heating heat exchangers.

All these systems can use aquifer thermal energy storage. The two-pipe system is recommended for a heat pump system which heats and cools. The four-pipe system is recommended for direct cooling and direct heating. The cooling and heating could be provided by central heating and cooling systems.

Pipe Materials - PVC

Metal pipes such as steel or cast iron present some problems on account of corrosion of dissimilar metals as well as economics (see Chapter V). It must be remembered that inhibitors are not permitted in the aquifer. Copper and brass can be eliminated directly on an economic basis. As a result, polyvinyl chloride (PVC) pipe is proposed as a distribution pipe. PVC pipe will:

    1. Carry the required pressure,
    2. Satisfy the temperature requirements in most applications,
    3. Minimize pipe cost and installation cost, and,
    4. Eliminate the possibility of dissimilar metal to metal
       corrosion from galvanic reactions.

Many city potable water distribution systems are beginning to install PVC pipe. The following specifications of polyvinyl chloride (PVC) piping is a summary of material from Piping Handbook (1).

Rigid, unplasticized polyvinyl chloride is a thermoplastic material; it is tough and exceptionally resistant to chemical attack. Plastic pipe is generally available in iron pipe sizes (IPS). Typical sizes and weights for small PVC sizes are given in Table 7.1.

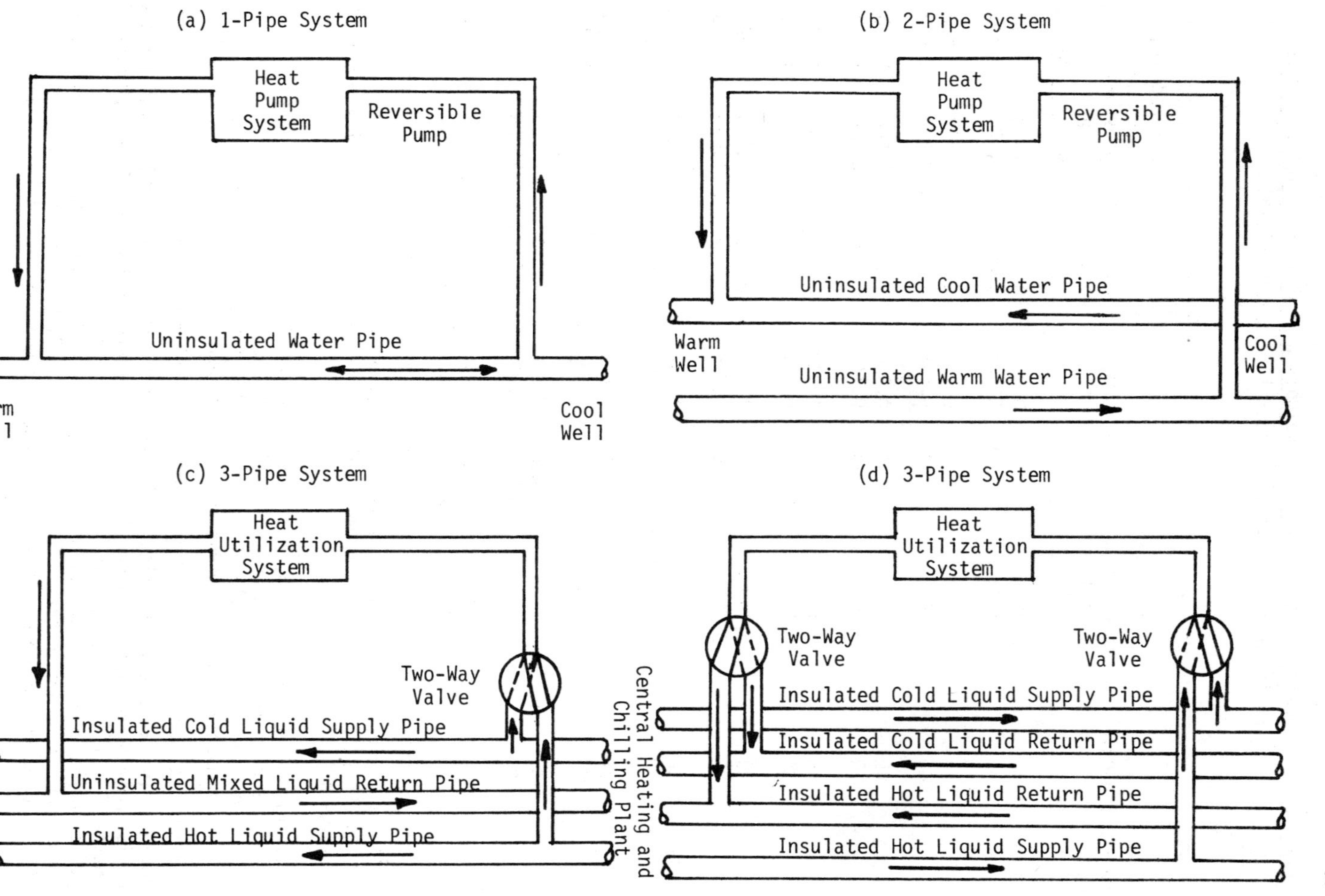

Fig. 7.3. Water distribution system piping configurations.

Table 7.1. Commercial Sizes (IPS) and Weights of Polyvinyl Chloride (PVC)
Pipe (Abstracted from ASTM Specification D1785-64T) (1)

Normal size in	Schedule	Wall thick-ness, **in	OD, in	ID, in	Theoreti-cal weight,* lb/ft	Calculated min bursting pressure psi Note 1	Note 2
1/4	40	0.088	0.540	0.364	0.076	2,490	1,950
	80	0.119	0.540	0.302	0.096	3,620	2,830
1/2	40	0.109	0.840	0.622	0.153	1,910	1,490
	80	0.147	0.840	0.546	0.195	2,720	2,120
3/4	40	0.113	1.050	0.824	0.203	1,540	1,210
	80	0.154	1.050	0.742	0.265	2,200	1,720
1	40	0.133	1.315	1.049	0.305	1,440	1,130
	80	0.179	1.315	0.957	0.385	2,020	1,580
1 1/4	40	0.140	1.660	1.380	0.409	1,180	920
	80	0.191	1.660	1.278	0.550	1,660	1,300
1 1/2	40	0.145	1.900	1.610	0.489	1,060	830
	80	0.200	1.900	1.500	0.653	1,510	1,180
2	40	0.154	2.375	2.067	0.640	800	690
	80	0.218	2.375	1.939	0.910	1,290	1,010
3	40	0.216	3.500	3.068	1.380	840	660
	80	0.300	3.500	2.900	1.845	1,200	940
4	40	0.237	4.500	4.026	1.965	710	560
	80	0.337	4.500	3.826	2.710	1,040	810

* These representative values are not specified in ASTM D1785-64T.
** Thicknesses listed are minimum values. Tolerance is generally -0 + 10 percent.

Note 1. Materials are PVC 1120, 1220, and 4116. A fiber stress of 6,400 psi was used in bursting pressure calculations.
Note 2. Materials are PVC 2112, 2116, and 2120. A fiber stress of 5,000 psi was used in bursting pressure calculations.

PVC pipe is extruded. Fittings, flanges, and valves are manufactured by the injection-molding method which results in high density and complete homogeneity of the material. Three types of PVC piping are available.

Type I in the past has generally been marketed as "normal-impact" grade. It is now produced to a hydrostatic design stress of 2,000 psi for water at 73.4°F (23°C). Two grades are recognized under the designations PVC 1120 and PVC 1220.

Type II has been marketed as "high-impact" grade. Type II, Grade 1 is produced to a hydrostatic design stress of 1,000 psi for water at 73.4°F (23° C) and is designated as PVC 2110. It is also produced to a hydrostatic design stress of 1,250 psi and is designated as PVC 2112; material produced to a hydrostatic design stress of 1,600 psi is designated as PVC 2116.

Type IV is produced to a hydrostatic design stress of 1,600 psi for water at 73.4°F (23° C). It is designated as PVC 4116.

Fittings and valves of Types I and II are readily available with threaded solvent weld (socket), and flanged ends. The two Type I PVC grades have greater strength properties over a wider temperature range. They can be used at temperatures up to 160°F and, in addition, at higher working pressures than can Type II PVC. These grades have superior chemical resistance throughout their temperature range.

Type I PVC grades should be specified in applications where greater strength, temperature resistance, or extreme chemical resistance is necessary. Type I is normally expected to be used in these systems. PVC in normally straw-colored without pigmentation. Industrial-process piping is usually colored dark gray for identification purposes but also may be produced as white, brown, red or any other color.

PVC pipe is covered by ASTM Standard D1785, tentative specifications for Poly Vinyl Chloride (PVC) Plastic Pipe. Voluntary industry standards have also been issued by the U.S. Department of Commerce: Commercial Standard CS256, Polyvinyl Chloride (PVC) Plastic Pipe (SDR-PR and Class T) and Commercial Standard CS207, Rigid Unplasticized Polyvinyl Chloride Pipe. The normal physical properties of PVC are summarized in Table 7.2.

Table 7.2. Normal Physical Properties of PVC Piping Materials (1)

Properties	Materials		ASTM Test No
	Type I	Type II	
Tensile strength, room temperature, psi	7,000	6,000	D638
Modulus of elasticity in tension, psi	415,000	350,000	
Flexural strength, psi	14,500	11,500	D790
Izod impact strength, ft-lb/in of specimen notched room temperature	0.5-1.0	10.0-19.0	D256
Izod impact stength, ft-lb/in of specimen unnotched room temperature	45	55	D256
Specific gravity	1.38	1.35	D792
Hardness, Shore "D"	78/82	76/80	D785
Heat distortion, temperature, F, at 264 psi	165	155	D648
Coefficient of thermal conductivity, BTU/(sec) ft F x 10^4	3.5	4.5	C177
Specific heat, cal/(g C)	0.25	0.25	
Coefficient of linear thermal expansion per C x 10^5	5	10	D696
Water absorption, percent in 24 hr at 25°C	0.07	0.07	D570

PVC piping is extensively used in highly corrosive applications involving acids, alkalies, salt solutions, alcohols, and many types of chemicals. PVC piping is also used in oil fields because it can carry sour crude oil to which PVC is chemically inert and because paraffin build-up is minimal on the smooth inside surfaces of this pipe. Other applications include saltwater disposal in oil fields and gas transmission service. PVC will handle most chemicals up to 150°F (66° C). Still other applications include the piping of cold water in industrial plants, because PVC is nontoxic and will not impart odor or taste to the water, and also vent piping for the removal of acid fumes and corrosive gases from industrial plants.

PVC pipe and fittings experience little or no physical deterioration when exposed to direct sunlight; sunlight causes deterioration in several other plastic piping materials. Like other plastic material, PVC will not produce sparks when struck. PVC is safe to use around explosive or flammable vapors, and it does not support combustion. Also water contaminants do not build up on the smooth walls of PVC pipe or fittings.

Because polyvinyl chloride, like most other plastic materials, is somewhat notch-sensitive, maximum allowable working pressures for threaded pipe are considerably lower than those of unthreaded pipe. The strength of polyvinyl chloride decreases as the operating temperature increases, and allowable working pressures must be decreased at higher temperatures. Working pressures at 75°F (24°C) are approximately 20 percent of the bursting pressures. When operating temperatures exceed 75°F (24°C), the maximum operating pressures for Type I pipe may be determined from Fig. 7.4.

Module Pressure Losses

By using the module system, pressure losses in the main water lines are minimized as a result of short travel distances. Calculated hydrostatic heads for designed systems can be less than 30 ft of water. Figures 7.5 and 7.6 give pressure losses for twenty home modules with homes separated by 100 ft and 200 ft, respectively, at various flow rates. A computer program was used for calculating pipe-pressure loads. Laminar flow is assumed in the piping which is reasonable for the smooth pipe with limited flow. The system assumes that each home receives the allotted water flow rate, thereby reducing the flow. This occurs at the given separation distances noted on the curve for two users at each cutoff. The total pressure loss is the pressure loss through the ten increments. Using pipe diameters of over four inches keeps the pressure losses at negligible values and allows flow both ways between the pipes. Change in elevation will not affect the pressure between pipes as they run parallel and adjacent.

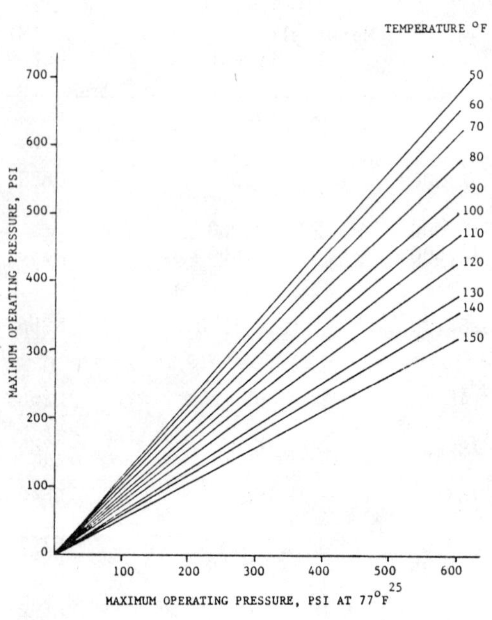

Fig. 7.4. Maximum operating pressure of normal-impact (type I) PVC at any temperature.

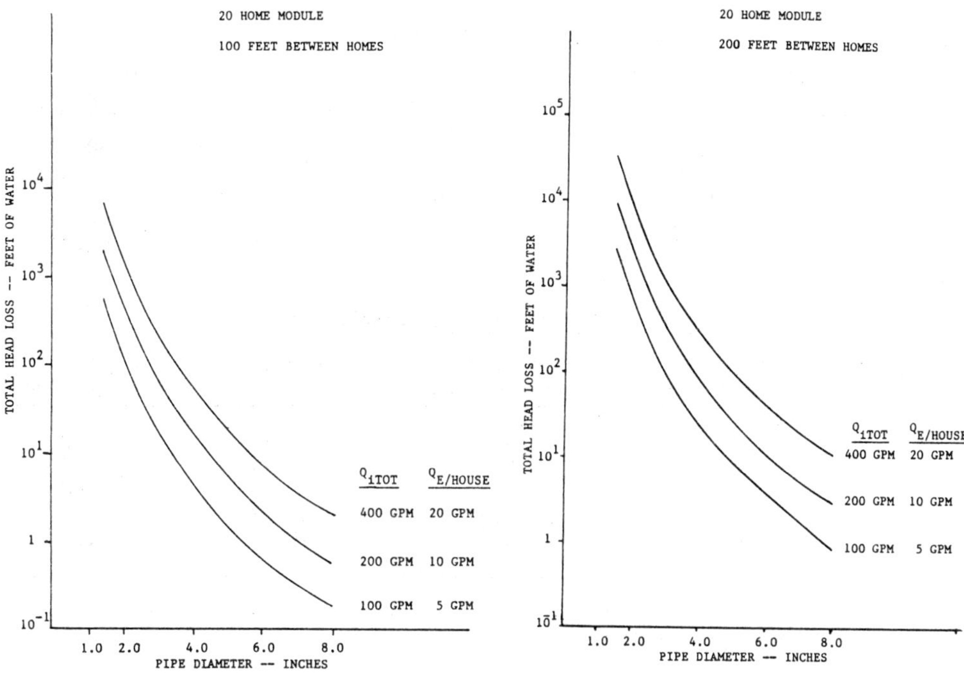

Fig. 7.5. Module total distribution pipe pressure loss.

Fig. 7.6. Module total distribution pipe pressure loss.

Connections between the main lines and the residential units are made with one and one-half inch or two-inch pipe. This will keep the pressure loss for flows up to 20 gpm below fifteen ft of water. A small pump can be used to circulate the water between the main water lines depending on the presure difference. Average maximum flow rate to a residential energy user is expected to be six gpm. Six-inch lines can be utilized as main water lines to provide adequate water for emergency purposes such as fire protection systems. It may also be noted the pressure loss in the main lines for six-inch pipe is well below one foot of water if two homes at opposite ends of a module heat and cool simultaneously (flow rate of ten gpm). The large lines also decrease the requirements on potable water pipelines. These are normally controlled by fire code standards.

Cost Analysis

An analysis has been made to give ballpark numbers on the cost of installing the various pipe configurations. The values for uninsulated and insulated pipes are given in Table 7.3. These numbers are for large jobs in excess of $100,000 in cost. They do not include costs for finished surfaces such as curbs and streets or for moving through other pipe or cable right-of-ways.

Table 7.3. Cost of Distribution Pipe System

Cost per Linear Foot

DIAMETER (INCHES)	UNINSULATED PIPE	ADD FOR INSULATION	TRENCHING EXCAVATION	BACKFILL
1 1/2	$1.70	$2.35*	$1.11**	$0.37***
2	2.05	2.40	1.11	0.37
4	3.20	3.95	1.11	0.37
6	4.60	7.25	1.11	0.37
8	7.25	8.80	1.11	0.37

Cost of Installed Pipe per Linear Foot

DIAMETER (INCHES)	1- PIPE UNINSULATED	2-PIPES UNINSULATED	3-PIPES 2 INS. 1 UNINS.	4-PIPES ALL INSULATED
1 1/2	$3.18	$ 4.88	$11.28	$17.68
2	3.53	5.58	12.43	19.28
4	4.68	7.88	18.98	30.08
6	6.08	10.68	29.78	48.88
8	8.75	15.98	40.83	65.68

All cost figures stated above were taken or derived from those given in
Means' Building Construction Cost Data (3).

The costs stated above are for Polyvinyl Chloride (PVC) pipe, class 160, SDR 26.

The pipe cost stated above includes material, installation, and subcontractor's
overhead and profit.

The trenching costs are stated for a four foot wide, three foot deep trench which
would accommodate any combination of pipes.

*   The costs for the 1 1/2" and 2" insulation is for 1/2" thick foam rubber
    sleeves and the cost of the 4" to 8" sizes is for 2" thick fiberglass
    insulation.

** This cost is for the use of a 1/2 cubic yard tractor backhoe.

*** This cost is based on using a dozer with up to a 300-foot hard distance and no
    compaction.

Water Distribution System Management

A significant open question in the adoption of integrated community energy
systems using thermal energy storage in aquifers relates to the ownership and
management of the water distribution systems. What type of organization is
required to design, finance, install, and maintain such systems?

The creation of a new utility, like those currently providing water, electricity, and natural gas, might be required. Some systems could be started without the development of a formal utility structure. The system would be initially designed, built, and financed by the developer of the new residential community. The management and maintenance of the system would be assumed by the homeowners association. Independent contractors might enter into agreements with homeowners associations for system maintenance. System maintenance fees could be collected by the homeowners association and distributed to the contractor on a fixed price or per-call basis.

As systems become more prevalent, a more formal utility structure will become necessary. That structure could be part of an existing water, electric, or gas utility or formed as a separate entity. The utility's ownership may be either private, as are many current supplies of electrical power and natural gas, or public like most water systems. Public ownership may offer advantages in terms of seeking investment funds via tax-free bonds. Croke, _et al_, (2) have described the financing of such systems. The inertia associated with municipal organizations may make a private utility company the most expeditious alternative.

The water distribution utility would have responsibilities beyond the servicing of the system. Some of its roles would include:

1. System planning
2. System maintenance
3. Emergency preparedness
4. Aquifer quality assurance
5. Heat balance control

The public health and welfare aspects of proper aquifer management demand careful monitoring and regulation of the water supply. The utility or a separate management board must be charged with that responsibility. With widespread adoption of the water source heat pump system concept, aquifer management will become a significant concern of local governments. A county water quality officer might be required.

With institutional application such as a military base, an individual in the base civil engineering office could be designated to manage the water distribution system. That individual would direct the on-base maintenance activities and coordinate aquifer management information with off-base officials and other aquifer users.

References

1. Crocker, S., _Piping Handbook_, McGraw-Hill, New York (1945).

2. Croke, K., Baum, J., and Rosenberg, R., _Municipal Financing of Integrated Community Energy Systems_, ANL/ICES-TM-3, (1977).

3. Means, R.S., _Building Construction Cost Data_, Robert Snow Means Co., Inc., Duxbury, Massachusetts (1978).

CHAPTER VIII

ENVIRONMENTAL ASPECTS

The effects of thermal energy storage on the aquifer environment are still vague. However, four major projects being financed by the U.S. Department of Energy are involved in investigations of the environment aspects. Presently the environmental effects may be broken into two categories:

1. State and federal environmental regulations which may apply to the aquifer systems in the detailed design and implementation phases.

2. Known and projected environmental impacts to provide a reference base for impact analysis. These include physical and biological impacts. They may be divided into catagories which affect the efficiency of the system and those which affect the human environment.

The state and federal regulations which cover well drilling and withdrawal of water from the aquifers are discussed in Chapter IV. These reguations vary widely from state to state. The National Water Well Association is compiling the regulations for each state on a DOE contract. Some environmental changes may occur in aquifers. Thermal cycling may affect interstitial minerals in aquifers, contaminate groundwater, and change groundwater temperature thereby accelerating biological growth. Promotion of the growth of iron algae is a possibility which could clog the aquifer passages. However, these effects are not expected to be of major consequence.

Physical Reactions

Precipitation of salts which could clog aquifers and piping could be of importance. Some salts increase in solubility as temperature drops (mainly carbonates), while others increase in solubility as temperature increases. The temperature range employed in most proposed systems is considered to be insignificant from a geochemical standpoint. The impact of changing solubility of salts in aquifers, mostly Ca and Mg salts, at the anticipated temperature ranges could be measured only in terms of geologic time.

The carbonate of calcium ($CaCO_3$) is the most predominant carbonate in nature and the primary constituent in limestone or dolomite aquifers. Also, carbonate may be a predominant substance in interstitial spaces in other aquifers, notably sandstones. These considerations should apply to most aquifers.

Calcium carbonate is dissolved by weak acids by the following:

$$CaCO_3 + H^+ \rightarrow Ca^{++} + HCO_3 \qquad\qquad (8\text{-}1)$$

and limestone dissolves by the equation:

$$CaCO_3 + H_2CO_3 \rightleftharpoons Ca^{++} + 2HCO^-_3 \qquad (8\text{-}2)$$

This is an equilibrium situation in which the reaction can go in either direction.
   The pH level is a controlling factor inasmuch as the hydrogen ion concentration is a function of equation 8-1. At low pH, the forward action results in solution, whereas at high pH the reverse action would occur leading to precipitation.
   A more dominant situation relates to the pressure of CO_2 above the solution. This determines the concentration of dissolved H_2CO_3 by the relationship:

$$H_2O + CO_2 \rightleftharpoons H_2CO_3 \qquad (8\text{-}3)$$

   Any process increasing the available CO_2 promotes dissolution; decreasing the amount of CO_2 promotes precipitation.
   Relation 8-3 has a bearing on the effects of temperature on the solubility of $CaCO_3$ in natural processes. The solubility of $CaCO_3$ in pure water "...decreases somewhat as the temperature rises." This is opposite to the behavior of most salts (1).
   Like any other gas, CO_2 is less soluble in hot water than in cold water. For instance, at cold ocean depths $CaCO_3$ is dissolved, whereas at warm surface temperatures there may be precipitation.
   It is considered that a temperature increase of possibly 25°F to 125°F (4°C to 52°C) in return water to an aquifer would not result in any change in the chemical makeup of the circulation water, nor would it affect the physical parameters of the aquifer. The pH of "pure" water is 7.0 but, if exposed to the atmosphere for a short time, will be near 5.7 as a result of the intake of CO_2 from the atmosphere (unless neutralized by other things). Any process in the pumping and circulation of water that affects pH might affect the solubility of both carbonates and salts. This should be minor. A solution to the problem is to prevent water-air interfaces in the system.
   Salts such as NaCl, K_2SO_4, and $Ba(NO_3)_2$ dissolve in water to form neutral solutions. Others such as iron sulfate may hydrolyze leaving an acidic solution. Granted these changes will occur in nature, the introduction of a minimal temperature escalation should have no significant effect on the nature of the aquifer or of the circulating water. Silica, the constituent of sandstones, is dissolved and forms a gel--in nature this may require thousands of years to be perceptible.
   Laboratory studies of the solubility of chemical substances have shown a requirement for high temperature increments over extended periods to realize any effects of temperature. In studies (2) involving the injection of liquid wastes into core samples under simulated reservoir conditions of temperature and pressure--i.e. 200°F and 600 psi, no difference in the permeability to flow was noted when compared with normal temperatures and pressures (surface conditions) (2). About ten different injection tests with different samples were made. The duration of the tests ranged up to some 100 days in certain instances.
   In summary, any change in aquifers under conditions of nominal increases in temperature of the return water--anything in the range of 50°F or even 100°F is expected to be negligible.

Biological Reactions

   Aquifer clogging from biological growth is minimal if a closed system (no contacts with air) is utilized. However, if cooling towers are used to chill the water, some problems may result. Industries using large quantities of natural water face problems of staining and clogging, caused by ferric hydroxide formed by microbes.

In most areas, for example, the principal iron bacteria in natural waters are Gallionella, a group existing in the majority of natural surface streams, ponds, small or large impounded waters and in soil, especially if the area is not disturbed by cultivation and nearby deposits of readily available nutrients and soluble organic compounds are readily available. These bacteria were found in 54 percent of 109 well water samples in Alabama collected from 35 of the 64 counties (3). In Tuscaloosa County, Alabama, six of nine well samples were positive. In addition to clogging of well screens and casing perforations and interfacing with flow, unpleasant odors and tastes develop and eventually the mass of bacteria provides nutrients for the growth of other microorganisms.

Although some iron bacteria require aeration and organic matter for growth, the species of Gallionella requires only small amounts of reduced iron (ferrous salts), a very low oxygen concentration, soluble carbonates, minimal quantities of minerals such as nitrates, sulfates, and phosphates and grow over a temperature range of 10^oF-30^oF (6^oC-17^oC). Problems in this area were anticipated by Davison (4) at Texas A&M where water is chilled in cooling ponds and returned to aquifers. The cooling pond precipitated the ferrous salts from the water; as a sand filter removed the particles, and no known problems of aquifer clogging have occurred. These bacteria attach to surfaces in moving water systems and, although dispersed throughout the system, appear to concentrate where local conditions are most favorable for growth which is often in pockets where the reduced-flow conditions occur in pipes. Detection is primarily microscopic although cultivation or concentration on membrane filters may be required for confirmation of new contamination. Aquifers are not considered suitable for bacterial growth because of the low concentration of available nutrients and the absorptive surfaces which filter out particles. Water returned to an aquifer, however, after going through a cooling tower has not been specifically examined. Some investigators have reported fouling by sulfate-reducing microbes and by slime-producing bacteria in open circulating water systems in refineries. The microbiological control program included total plate counts (a cultivation method for quantification) as well as specific identification of the undesirable species, Desulfovibrio desulfuricans (an anaerobe) and the slime formers which are essentially aerobic. Additional non-microbial detection methods were used. These results were used to determine water treatment. Results were not sufficiently conclusive to determine if some of the microbial work could predict problems before fouling occurred (5). These studies included considerations of:

1. Gross and microscopic particulate matter
2. Turbidity
3. Dissolved CO_2 and/or carbonates
4. Total organic matter or carbon or BOD/COD
5. Quantitative microbial determination
6. Specific qualitative tests for microbes as discussed later.

The above tests should be done at sufficient sites to determine normals and trends. This will then allow predictions to be made of impending problems in actual operation.

The second concern, any effects upon man's health, is important, but no previous investigations have been reported in this field. However, a recent report from the Center for Disease Control of USPHS, Atlanta, Georgia, found the Legionnaires's Disease bacterium in cooling towers of air conditioning systems as well as in many other natural sources (6). The present methodology for detection and identification requires concentration of the water samples by filtration and centrifugation and subsequent fluorescent antibody tests. The flourescent antibody test is a very specific identification by special microscopic examinations and with antiserum and reagents specified and supplied by the Center for Disease Control (7).

Some preliminary tests done in the Department of Microbiology at The University of Alabama on cooling-tower waters showed the water to be grossly contaminated with bacteria. Specific identification was not attempted. Screening tests for a few microbes which can cause human infections if inhaled should be considered. These should establish both methodology and predictive tests for protection of man's environment. The use of inhibitors and numerous other chemicals is prohibited in shallow aquifers according to the EPA <u>Interim Primary Drinking Water Regulations-Control of Organic Chemical Contaminants in Drinking Water</u>. The use of chlorine for biological control should not be a problem.

Water Table Impact

The water table is not significantly affected inasmuch as the same quantity of water is injected as is withdrawn. If cooling towers or ponds are utilized, water may be discarded to maintain thermal equilibrium. This would use some water from the aquifers, thereby limiting use of these types of systems in some parts of the country.

Summary

As may be noted, the final environmental aspects of aquifer use are still unknown but appreciable progress has been made and extensive studies in this area are being performed under major U.S. Government sponsorship. References 8 to 14 include studies of various problems. The reports and papers note numerous applications of aquifer use, including wastewater injection. Reference 13 gives the environmental protective steps required for well construction. No major problems are anticipated in using aquifers for thermal energy storage. For the most part, overall environmental changes are expected to be positive; minor adverse effects can be dealt with or justified in light of the beneficial aspects of the system performance.

References

1. Krauskopf, K.B., <u>Introduction to Geochemistry</u>, McGraw-Hill Book Company, (1967).

2. Grubbs, D.M., Haynes, C.D., Hughes, T.H., and Stow, S.H., "Compatibility of Subsurface Reservoirs with Injected Liquid Wastes," NRC Report 721, The University of Alabama (June, 1972).

3. Valkenbury, N., Christian, R., and Green, Margaret M., <u>Occurrence of Iron Bacteria in Groundwater Supplies in Alabama</u>, Circular 96. Geological Survey of Alabama, (1975).

4. Davison, R.R., Harris, W.B., and Reddel, D.L., "Texas A & M Cold Water Aquifer Storage Project," in <u>Proceedings of Solar Energy Storage Options</u>, Trinity University, (1979).

5. Foster, B., and Becher, J., <u>Control of Problems in Refinery Open Circulating Cooling Water Systems</u>, Unpublished report. ASME Winter Annual Meeting, New York City, (1979).

6. Click, T.H., Gregg, M.B., Berman, B., et al, "1978 Pontian fever: An Epidemic of Unknown Etiology in a Health Department. I. Clinical and Epidemiologic Aspects." <u>Am. J. Epidemiol</u>. 107: 149-160 (1979).

7. Jones, G.L., and Hebert, G.A., <u>"Legoionnaires," The Disease, the Bacterium and Methodology</u>, Center for Disease Control, Atlanta, Ga., (1978).

8. A Manual of Laws, Regulations, and Institutions for Control of Ground Water Pollution, NTIS No. EPA-440/9-76-006, (1976).

9. Proceedings of the Third National Ground Water Quality Symposium, NTIS No. EPA-600/9-77-014, (1977).

10. Impact of Abandoned Wells on Ground Water, NTIS, No. EPA-600/3-77-095, (1977).

11. Water Conditioning Technical Manual, Meyers & Bro. Co. Ashland, Ohio, (1968).

12. Manual of Water Well Constrution Practices, NTIS No. EPA-570/9-75-001, (1975).

13. An Introduction to the Technology of Subsurface Wastewater Injection, NTIS No. EPA-600/2-77-240, (1977).

CHAPTER IX

HEAT PUMP COOLING AND HEATING SYSTEMS

The mating of heat pumps and aquifer thermal energy storage is like combining a horse and carriage. The heat pump is a prime system for removing or adding energy to an aquifer system (1). Numerous varieties of small (1/2 ton) to large (over 2,000 tons) water source heat pumps are on the market. These heat pumps absorb or reject thermal energy from water.

The basic varieties of suggested heat pump systems are illustrated in Fig. 9.1. Individual homes may have a small individual heat pump. A larger building may

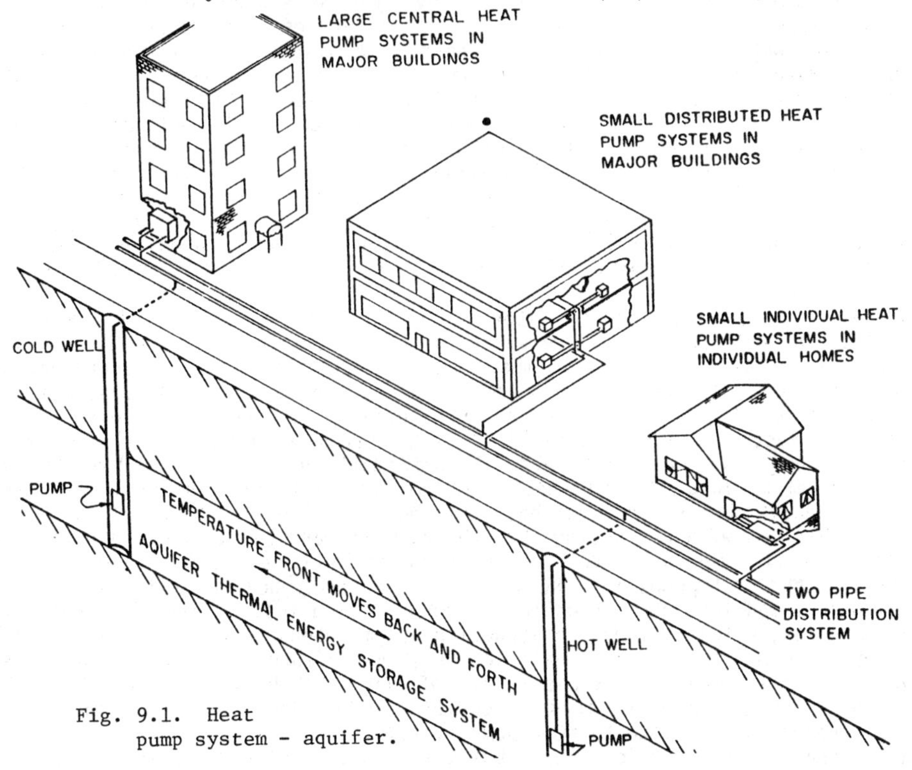

Fig. 9.1. Heat pump system – aquifer.

have a distributed system of heat pumps which heat a large area or a small single room. The larger building could have a central unit which heats and cools individual areas and rooms. A residential or commercial section could also have a central heat pump system which heats and cools homes and businesses. The heat pump upgrades energy in the aquifer making it effective for heating and cooling.

Basic System Description

The heat pump system described here utilizes annual thermal energy storage to raise the performance of individual heat pump systems. The system stores heat rejected by air conditioning and uses the energy for heating. As energy is used for heating, a heat sink is created for cooling.

A large capacity storage system is provided by nature in the aquifer. Access to the storage system is by a pair of wells. For the heating process, warm water, 80°F for the basic system, is withdrawn from storage and upgraded by a heat pump to 120°F for actual heating. In the process the water is cooled to 60°F and injected into the second well. A thermal energy front moves through the aquifer between the two wells. The quantities of water withdrawn and injected are equal. The net water flow from the aquifer is zero. For cooling the process is reversed. Cool water, at 60°F, is withdrawn from the aquifer and absorbs thermal energy to provide cooling. The warm water at 80°F is injected into the aquifer. As in heating, the net water usage is zero. During this process, a temperature front moves in the opposite direction between the pair of wells. Figure 9.1 represents such a system.

The heat pump system and thermal energy storage system are connected by a two-pipe water distribution system. One pipe carries warm water, and one pipe carries cold water. For heating, the heat pump system with an individual pump withdraws water from the warm pipe and rejects water to the cool pipe after withdrawing thermal energy. For cooling, the process is reversed. Heating and cooling can operate simultaneously. Pumps submerged in the hot and cold wells keep the respective pipelines filled with warm and cool water. Local aquifers are utilized so water flow is limited to short distances. No insulation is required because the pipelines are buried and operated near below-ground temperatures. To minimize pipe cost and installation cost, PVC (Polyvinyl Chloride) pipe is recommended for most installations. This pipe is being used in many water distribution systems. Iron or steel pipes are required for very high flow rates in major downtown areas.

To make the heat pump aquifer storage system feasible, a community system is required. The well pair costs are shared by several users. For technical reasons the thermal energy storage system must be above a minimum size to assure adequate storage and must be operated by a community body. The distribution system must be developed, constructed, maintained, and controlled by the community body.

Heat pump systems are purchased, installed, and maintained by building owners. These heat pump systems could range from a small heat pump in a home or major building room to large central heat pump systems utilized for large buildings. For simultaneous heating and cooling, warm and cool water can be exchanged between heat pump systems without passing through the wells and the aquifer thermal energy storage system. The well pair or a group of well pairs, an aquifer storage system, and the heat pumps in a section of a community heating and cooling system are called a module. By connecting a group of modules, reliability and system size increases, allowing lower construction and maintenance costs and a reasonably sized controlling body. For example, a pump failure could be repaired when convenient rather than during the middle of the night because a pump in another module will carry the extra load. Modules can be added one at a time to allow orderly expansion of community heat pump systems.

Rather than using a critical fuel, the system uses electrical power which may be generated by coal, nuclear, oil, gas, or hydro energy. A thermal-energy operated water-source heat pump employing absorption can be fitted into the system if waste energy is available.

Potential Applications

The proposed system has applications for any heating and cooling system located near an aquifer with sufficient yield capacity. Over 60 percent of the land area and over 75 percent of the population in the continental United States fall into this category. This includes downtown areas, commercial neighborhoods, and residential neighborhoods. The denser-use areas normally increase the feasibility. Other than the southern half of Florida and parts of California (mild year-around temperatures), the system can be applied with appreciable energy savings and economic savings, especially as energy costs increase.

Using only technological limits, the system could be incorporated by the year 2000 into

72,000,000	Households
14,600,000	Businesses.

Expected Performance

The expected performance of the proposed heat pump systems is evaluated by application to a typical home in a residential neighborhood. The performance of the proposed system is compared to present systems, including resistance heating, oil and gas heat, and air-source heat pumps. The comparison is made in heating and cooling system coefficient of performance and total energy utilized, namely the source energy supplies. The proposed system shows an advantage over standard present day heating and cooling systems not only in performance, but also in the flexibility of any available energy source. These include hydro, coal, nuclear, gas, oil and solar.

The energy balance diagram of the proposed system and present heating and cooling systems is shown in Fig. 9.2. The new development is the economical annual thermal energy storage provided by nature. Energy rejected during cooling periods is stored and used during heating periods. The sequence is shown in Part A of Fig. 9.2. This shows the energy going into the power plant being converted to work for use in the heat pump. The heat pump draws and rejects heat to thermal energy storage in the aquifer for heating and cooling cycles respectively. The normal air-to-air heat pump draws and rejects heat from the environment. The energy in the storage system is more efficiently used in heat pump installations. For example, when a normal system is pumping heat from $0°F$ at ambient conditions to $70°F$ at the controlled environment conditions, the proposed system is pumping heat from $80°F$ in storage to the $70°F$ controlled environment. These conditions are much more favorable and require less work input. The furnace-type heating as shown in Part C also has waste heat losses. The power plant is included in the schematic to relate performance for all systems on the basis of paid-for-energy -- fuel into the system.

The proposed systems are first evaluated by coefficient of performance (COP) at the point of utilization which is defined as:

$$COP_{Cooling} = \frac{\text{Energy Removed from Controlled Environment}}{\text{Work Required to Remove Energy}} \qquad (9\text{-}1)$$

$$COP_{Heating} = \frac{\text{Energy Added to Controlled Environment}}{\text{Work Required to Add Energy}} \qquad (9\text{-}2)$$

$$\text{Power Plant Eff.} = \frac{\text{Work Out (Electrical Energy)}}{\text{Thermal Energy Input}} \qquad (9\text{-}3)$$

112

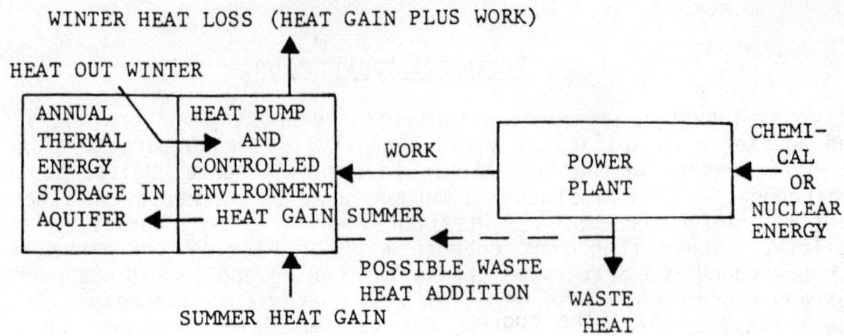

A.   PROPOSED HEAT PUMP SYSTEM

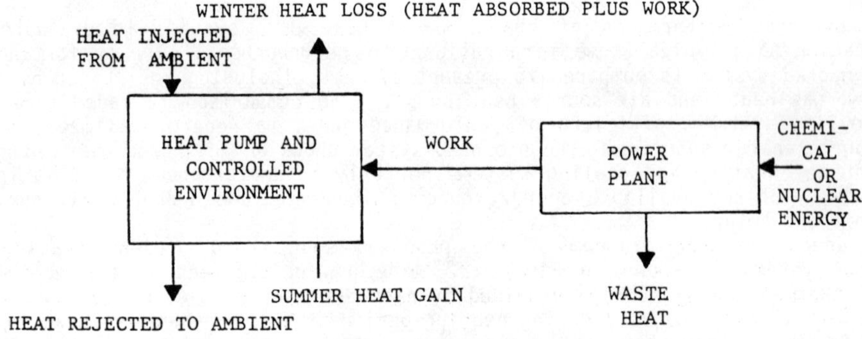

B.   AIR SOURCE HEAT PUMP SYSTEM

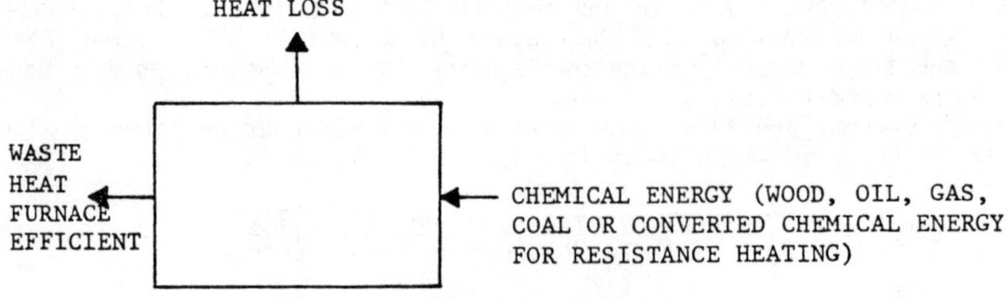

C.   DIRECT HEATING (FURNACE)

Fig. 9.2. Energy balance diagram.

However, to develop a physical feel for the increased performance, the cycles will be considered as ideal. The Carnot efficiency, the maximum possible efficiency, for heat pumps as a function of absolute temperatures are:

$$COP_{Cooling} \quad = \quad T_L/(T_H-T_L) \tag{9-4}$$

$$COP_{Heating} \quad = \quad T_H/(T_H-T_L) \tag{9-5}$$

$$Power\ Plant\ Eff. \quad = \quad (T_H-T_L)/T_H \tag{9-6}$$

The main factor in these expressions is the temperature difference in the denominator. A small change in this difference makes a large change in performance. Using annual aquifer storage in parallel with heat pumps decreases this energy difference. The warm water rejected from the cooling cycle is stored and used for heating. The cool water rejected from the heating cycle is stored and used for cooling.

To show the potential of these systems, 70°F (530°R) water is used to provide a 50°F (510°R) temperature and 120°F (580°R) heating temperature. The ideal coefficients of performance according to Carnot cycles (absolute temperatures) are:

$$COP_{Cooling} \quad = \quad 510°R/(530°R-510°R) \quad = \quad 25.5$$

$$COP_{Heating} \quad = \quad 580°R/(580°R-530°R) \quad = \quad 11.6$$

These values represent possible performance at the site using ideal (electrical) energy input or work. To convert these values to resource energy the power plant efficiency and energy distribution losses must be taken into account.

The ideal efficiency is again the Carnot cycle. Assuming energy from chemical fuel has a 3500°R potential and a 77°F (537°R) rejection temperature, the ideal power plant efficiency is:

$$Power\ Plant\ Eff. \quad = \quad (3500°R-537°R)/3500°R \quad = \quad 0.847$$

Overall system performance is COP times power plant efficiency or overall performance Carnot capability for

$$Cooling \quad = \quad 25.5\ X\ 0.847 \quad = \quad 21.6$$

$$Heating \quad = \quad 11.6\ X\ 0.847 \quad = \quad 9.83$$

These numbers represent the cooling or heating energy available for a unit of source energy. The numbers are very sensitive to the temperature differences. The proposed systems and air source heat pump system move in the direction of ideal performances. Using high quality energy for direct heating is a waste as overall performance is limited to less than 1.0. The system performance is maximum, air-source heat pumps rank second, and hydrocarbon and resistance heating rank last.

Numerous studies have been made using the Earth and other types of storage to minimize these temperature differences (2,3). This has been true not only in the United States, but also in Europe (4,5). In many of the systems, the heat transfer is by pipes in the Earth. In this text nature is allowed to provide the heat transfer mechanism, that is, porous underground rock formations through which water can flow to transfer the thermal energy. A large, annual storage system is developed for a community energy system.

The coefficients of performance are applied directly to the heating and cooling sequence in the basic and alternative systems developed in this chapter. For comparison purposes, the listed performance for approximately three-ton water source heat pumps is utilized. The high performance unit of ARI listing (6) is used for the water-source heat pump. The basic performance is corrected for changes in water temperature, air temperature, and humidity using detailed data from American Air Filter. General comparisons with data from other manufacturers showed these corrections are valid. The high-performance unit is used since a water source heat pump is designed to a fixed temperature range and should have a higher performance than those designed for a wide range.

The variations with performances of water-source heat pumps in the basic system and alternative systems follow. Compressor and fan power are included in the ARI listings (6). A conservative estimate of 300 watts for water pump power is utilized. All these systems use the two-pipe water distribution systems. One pipe contains relatively warm water and one pipe contains relatively cold water. The systems are:

1. Basic: This system supplies water directly to the heat pump for cooling at 60°F and for heating at 80°F.   The high performance water-source heat pump operating characteristics are given in Table 9.1.

TABLE 9.1. System Performance of Water Source Heat Pump in Basic System

Water temperature 60°F(15.5°C) for cooling; 80°F(26.7°C) for heating

Heating

Entering water temperature °F (°C)	Capacity[1] (KW)	Power[2] (KW)	COP[3]
59 (15.0)	8.00	2.84	2.82
70 (21.1)	9.50	3.22	2.96
81 (27.2)	10.99	3.52	3.13
90 (32.2)	12.51	3.95	3.16

Cooling 67°F(19.4°C) WBT

Entering water temperature °F (°C)	Capacity[1] (KW)	Power[2] (KW)	COP[3]
64 (17.8)	11.23	3.33	3.37
75 (29.9)	10.81	3.43	3.15
84 (28.9)	10.40	3.60	2.89
95 (35.0)	9.99	3.73	2.68

(1) Based on ARI listing (High Performance Unit)
(2) Power for compressor, fan, and water pump
(3) COP = capacity (KW)/input to compressor + fan + 0.3 KW for water pumping

2. Alternative I: This alternative uses the same water temperatures as the basic system. However, a theoretical 50 percent-efficient heat exchanger is placed between the air flow and water flow (Fig. 9.3). As a result, some direct heat exchange occurs which does not require heat pump work. This becomes particularly effective with the new thermostat settings of 78°F (60°F cooling water) for cooling and 65°F (80°F heating water) for heating. Performance shows a small finite step increase for both cooling and heating. Performance for water-source heat pumps is given in Table 9.2 for this alternative.

TABLE 9.2. System Performance for Modified Water Source Heat Pump with Water to Air Heat Exchanger – Alternative I

Water temperature 60°F(15.5°C) for cooling; 80°F(26.7°C) for heating

Heating

IDBT[4] °F (°C)	Pump capacity (KW)	Water coil capacity (KW)	Total capacity (KW)	Power[2] (KW)	COP[3]
61(16.1)	9.50	4.10	13.60	3.22	4.22
64(17.8)	9.88	3.05	12.92	3.29	3.92
70(21.1)	11.75	2.02	13.77	3.74	3.69

Cooling 67°F(19.4°C) WBT

IDBT[4] °F (°C)	Pump capacity (KW)	Water coil capacity (KW)	Total capacity (KW)	Power[2] (KW)	COP[3]
70(21.1)	11.23	2.02	13.25	3.33	3.98
75(29.9)	11.14	3.05	14.19	3.35	4.23
81(27.2)	11.02	4.04	15.06	3.38	4.46
84(28.9)	10.93	5.07	16.00	3.40	4.71

(1) Based on ARI listing (High Performance Unit)
(2) Power for compressor, fan, and water pump
(3) COP = capacity (KW)/input to compressor + fan + 0.3 KW for water pumping
(4) Indoor dry bulb temperature

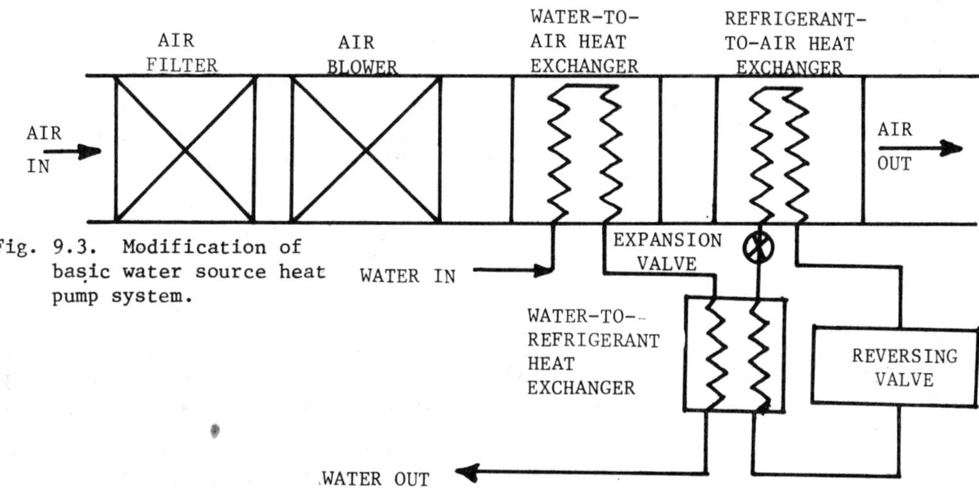

Fig. 9.3. Modification of basic water source heat pump system.

3. Alternative II: This alternative uses a low water temperature range of 40°F to 60°F. As a result, the cooling is accomplished by direct heat exchange without heat pump work. The performance for cooling more than doubles. However, due to a lower temperature water for heating, the heating performance shows a small decrease in performance. This system is similar to the ice-making heat pump system. This system uses 40°F water in a nature-provided aquifer rather than in a man-produced storage system for ice. This system is "free cooling." The performance is listed in Table 9.3.

TABLE 9.3. System Performance for Modified Water Source Heat Pump A with Water to Air Heat Exchanger – Alternative II

Water Temperature for Heating – 60°F (15.5°C)

IDBT[2] °F (°C)	Air flow (cfm)	Water flow (gpm)	Capacity[1] (KW)	Power[3] (KW)	COP[4]
61 (16)	1000	4	8.32	2.84	2.93
64 (18)	1000	4	8.18	2.84	2.88
70 (21)	1000	4	8.00	2.84	2.82

Water Temperature for Cooling – 67°F (19.4°C) WBT

IDBT[5] °F (°C)	Air flow (m³/min)	Water flow (gpm)	Capacity[1] (KW)	Power[6] (KW)	COP[4]
61 (16.1)	28.3	2.42	5.92	.9	6.51
64 (17.8)	28.3	3.91	10.29	.9	11.0
66 (18.9)	28.3	4.88	13.28	.9	14.1
70 (21.1)	28.3	6.00	16.53	.9	18.0

(1) Based on ARI listing (High Performance Unit)
(2) Indoor dry bulb temperature
(3) Power for compressor + fan + 0.3 KW for water pumping
(4) COP = capacity (KW) input power
(5) Indoor wet bulb temperature
(6) Power input for fan = 0.3 KW for water pumping

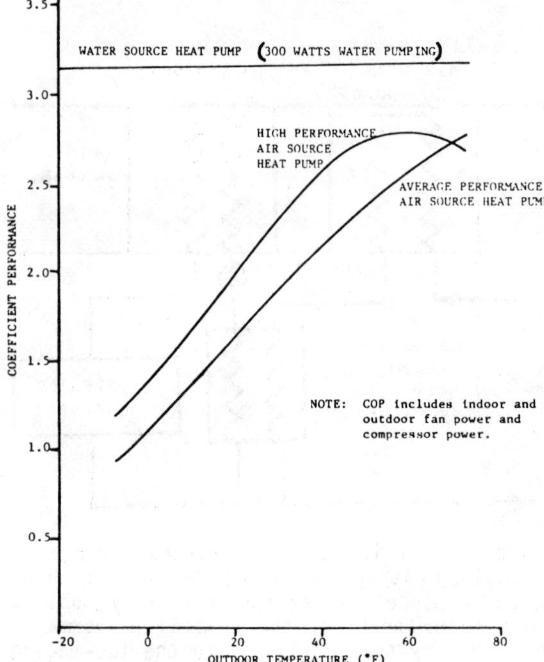

Fig. 9.4. Heat pump heating performance comparison.

4. Alternative III: This alternative is the opposite of Alternative II as the higher temperature range of 80°F to 100°F is used for water stored in the aquifer. In this case the hot water line must be insulated. The heating is by direct heat exchange, and the heating performance more than doubles. However, the cooling performance shows a decrease due to the warmer water. This system is "free heating." The performance for water-source heat pump A is given in Table 9.4.

The heat pumps and heat exchangers for these systems in the operating temperature ranges are available as off-the-shelf items. The matching of components for the upper range temperature in large-sized equipment must be analyzed for component and compressor capabilities in the entire pressure range.

Figures 9.4 and 9.5 show a major advantage of the proposed system performance versus the air source heat pump's performance. Neither the coefficient of performance nor

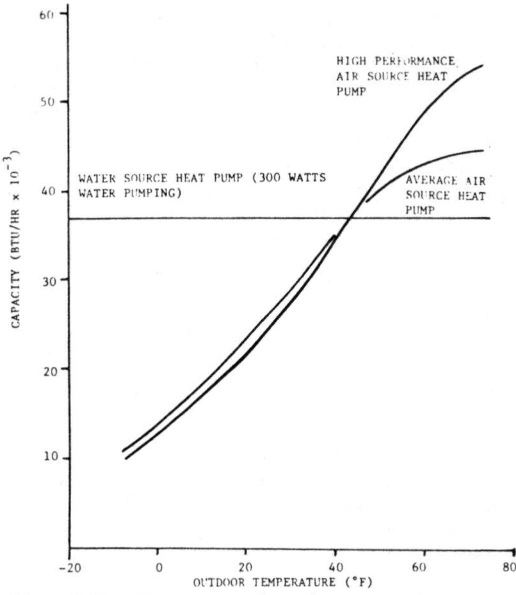

Fig. 9.5. Heat pump heating capacity comparison.

the capacity are affected by ambient conditions in the proposed system. The air source heat pumps at 0°F have lost 80% of their 60°F capacity and 60 percent of their performance.

The performance values for the proposed system and alternatives have higher potential in the future. The average ARI-listed water source heat pump performance data have potential for improvement especially in small units. New compressors of higher efficiency have been introduced to the market in the past few years. Basic data from Tecumseh support this fact (7). In addition, motor efficiencies have improved and heat-exchanger areas have increased. Estimates for these changes will increase water source heat pump performance by 20 to 30 percent. This means the proposed system and alternatives have potential performance increases of the same magnitude.

TABLE 9.4. System Performance for Modified Water Source Heat Pump with Water to Air Heat Exchanger – Alternative III

Water temperature 80°F(26.7°C) for cooling; 100°F(37.8°C) for heating

Heating

IDBT[1] °F (°C)	Air flow (ccfm)	Water flow (m^3/hr)	Capacity[2] (KW)	Power[3] (KW)	COP[4]
61 (16)	2000	4.79	15.12	1.05	14.40
64 (18)	2000	4.17	13.39	1.05	12.75
70 (21)	2000	3.60	11.64	1.05	11.09

Cooling 67°F(19.4°C) WBT

IWBT[5] °F (°C)	Air flow (ccfm)	Water flow (m^3/hr)	Capacity[6] (KW)	Power[7] (KW)	COP[4]
61 (16.1)	1000	4.48	8.47	3.6	2.35
64 (17.8)	1000	4.61	9.09	3.7	2.46
66 (18.9)	1000	4.79	9.79	3.8	2.58
70 (21.1)	1000	5.01	10.11	3.9	2.59

(1) Indoor dry bulb temperature
(2) Water coil + fan work
(3) Power input for fan + 0.3 KW for water pumping
(4) COP = capacity/input power
(5) Indoor wet bulb temperature
(6) Based on ARI listing (High Performance Unit)
(7) Power for compressor + fan + 0.3 KW for water pumping

118

Systems by WESCORP (8) have incorporated the new compressors, new motors, and improved heat exchangers. The basic performance in these references is over 40 percent higher than performance of present average ARI listing. However, capital costs are over 100 percent higher. At standard conditions without the work for water pumping, the COP for heating is over 4.2 and the COP for cooling is over 3.5 for units in the five to ten ton range. Variation in performance with water temperature is smaller than for the standard water-source heat pumps. Placing these units in the proposed system with the work for water pumping and applying the performance increase could result in future performance 20 percent greater than the present evaluations. Note that there is some question on the testing conditions for the WESCORP performance data. However, a performance evaluation is required to verify the guarantee for each installation. These evaluations have matched predicted values.

In addition, heat exchangers are available for these units to heat hot water. The superheated refrigerant leaving the compressor is used for heating the water. The result is that hot water is basically energy-free during cooling periods and will be heated with a COP of over 4.2 during heating periods.

Accurate capital costs on heat pumps are hard to obtain; however, general list price data by WESCORP indicate a capital cost of over two times the cost of standard water source heat pumps. With large usage and competition these prices are expected to decrease. Other manufacturers, including American Air Filter, are expected to have similar performance units within the next two years. Standard price and installation costs are estimated in Christian (9,10) for the smaller heat pumps. It should be noted that water source heat pumps are much more economical on a capital cost and installation basis.

Table 9.5 gives the performance of large reciprocating water source heat pumps. The performance is slightly higher than that of the smaller units. The percent

TABLE 9.5. Heating and Cooling Performance for Large*Roof Mounted Water Source Heat Pumps

System	Entering water temp. °F(°C)	Water flow heat/ cool gpm	Heating capacity MBTUH	Input[1] power KW	Heating[2] COP	Cooling capacity MBTUH	Input[1] power KW	Cooling[2] COP
RCRM - 24@	60(15.6)		249	21.9	3.33	310	25.4	3.58
8,000 cfm	70(21.1)	30	273	23.3	3.45	299	27.0	3.24
226 m³/min	80(26.7)	40	298	24.8	3.52	284	28.8	2.89
	90(32.2)		320	26.6	3.53	267	30.4	2.58
RCRM - 33@	60(15.6)		331	30.6	3.17	415	33.5	3.64
10,000 cfm	70(21.1)	42	363	32.4	3.28	399	35.5	3.29
	80(26.7)	56	396	34.6	3.36	380	38.0	2.93
283 m³/min	90(32.2)		425	37.1	3.36	356	40.1	2.61
RCRM - 42@	60(15.6)		450	41.2	3.21	555	46.0	3.54
12,000 cfm	70(21.1)	53	495	43.7	3.32	535	48.9	3.20
340 m³/min	80(26.7)	71	539	46.6	3.39	508	52.3	2.85
	90(32.2)		579	49.9	3.40	477	55.2	2.54

*Based on specifications from American Air Filter catalog AHU-1-121-JAN-02 for heating temperature of 70°F and cooling wet bulb temperature 67°F, with water pump power rated at 40w/gpm.

[1] Power for compressor, fan, and water pump.

[2] COP = Capacity (Kw)/Kw (compressor + fan + water pumping)

change in performance and capacity as a function of temperature are extrapolated to the large centrifugal and screw-type compressors, since general performance data for these units over wide temperature ranges have not been available. This assumption is based on the compressor pressure ratios being direct functions of temperature. The work input to the units, assuming compressor efficiency is constant, is a direct function of pressure ratio. Tables 9.6 and 9.7 give the predicted increases in performance and capacity for the large centrifugal and screw-type compressor units when utilized in the proposed basic system with improved water source temperatures. The performance of these large units does not include internal circulation or source water pumping. Nevertheless the performance is well above that of the smaller units. Performance and capacity show increases with a corresponding decrease in capital cost for identical output. An additional capital cost decrease is the elimination or decrease in size of water tower requirements.

A major benefit is a reduction in power plant peak loads. Using Alternative II in parts of the South and the East Coast will help eliminate peak summer loads. Matching loads to system alternatives, to types of buildings, and to power supplies is a performance possibility. The ratio of peak loads with Alternative II to air source heat pump for air conditioning is proportional to the ratio of COP's or a decrease of over 60 percent. Integrating some of these systems into an electrical system decreases power plant peak loads. A second major benefit is that oil and gas are not necessary. The power plants can operate on coal or nuclear energy, saving critical fuels.

The basic system and alternatives are compared to the standard heating and cooling systems. These include gas and oil heating, resistance heating, standard air-source air conditioning, and standard air-source heat pumps. Oil and gas efficiencies are indicated at 62 percent which includes performance and combustion efficiency. The tables show the total energy in electrical units where applicable or as fuel heating values for furnaces. These quantities are at the point of comsumption. Also, the electrical units are changed to total heating values for fuel entering the power plant. This allows energy consumption comparisons at the point of raw energy usage. A 40 percent efficient plant is utilized which represents a highly efficient fossil fuel plant.

For the vast majority of the country, with the exception of the lower half of Florida and part of the western coastal area, the system shows a sizable increase in performance. A summary of expected performance nationally is shown in Figs. 9.6 and 9.7. The numbers for the water source heat pump are for high performance units presently being sold. The data for the air to air heat pumps are averages obtained from the analysis of a 24 city study. The COPs utilized for the basic proposed system are 4.0 for heating and 3.5 for cooling. The alternative systems have higher performance in most cases. The values used for the air-source heat pump systems are 1.8 for heating and 2.8 cooling. These correspond to the high performance units presently on the market.

Heating to Cooling Load Ratio

The heating capacity of an annual heat pump system or heat pump-direct cooling or direct heating system is greater than the cooling capacity. This occurs because both the heating and cooling modes external energy is added to the system. This external energy goes into heating output.

The ratio of $Q_{Heating}$ to $Q_{Cooling}$ can be determined from the definitions of heating and cooling coefficients of performance. By matching terms the ratio is

$$Q_{Heating}/Q_{Cooling} = (COP_C+1)/COP_C \times COP_H/(COP_H-1) \qquad (9-7)$$

When both the cooling and heating coefficients of performance equal three, the ratio is:

TABLE 9.6. Screw Type Chillers Performance and Cost Data*

System A Capacity Tons	System B Capacity Tons	System A COP	System B COP	System A Capital Cost $	System B Capital Cost $	System A Inst. Cost **	System B Inst. Cost **	System A Total Installed Cost $	System B Total Installed Cost $
					Open Chillers				
500	454.44	4.70	5.97	54,000.00	50,708.11	3,825.00	3,540.84	57,825.00	54,248.95
750	681.82	4.70	5.97	70,569.00	66,266.74	5,312.09	5,312.09	75,881.09	71,184.17
					Hermetic chillers				
100	90.91	4.18	5.31	18,666.96	17,529.00	1,038.64	961.48	19,705.60	18,490.48
200	181.82	4.22	5.36	29,495.34	27,697.28	1,820.96	1,685.68	31,316.30	29,382.96
500	454.55	4.33	5.50	54,000.00	66,266.74	5,312.09	4,917.43	75,881.09	71,184.17

*System A Capacity is the capacity required for the building. System B is the standard capacity requirement which gives the identical output as System A operating with 60°F condenser water rather than the 80°F water produced by cooling towers.

Performance and capacity variations with temperature taken from American Air Filter Catalog AHU-1-121-JAN-02.

Performance (COP) is based on chiller power input. Values do not include water flow pumping or air flow distribution.

Unit cost data and installation man-hours taken from J. E. Christian, Central Cooling – Compressive Chillers, ANL/CES/TE78-2, 1978.

**$17.00 per man-hour is assumed.

TABLE 9.7. Centrifugal Chiller Performance and Cost Analysis[*]

System A Capacity Tons	System B Capacity Tons	System A COP	System B COP	System A Capital Cost ($)	System B Capital Cost ($)	System A Inst. Cost **	System B Inst. Cost **	System A Total Installed Cost ($)	System B Total Installed Cost ($)
					Open Chillers				
90	81.32	4.16	5.28	17,413.00	16,351.62	953.68	882.84	18,366.68	17,234.46
500	454.55	4.70	5.97	54,000.00	50,708.11	3,825.00	3,540.84	57,825.00	54,248.95
1000	909.09	4.90	6.22	85,324.46	80,122.43	6,706.79	6,207.79	92,030.49	86,330.22
1250	1,136.36	4.81	6.11	98,863.09	92,835.50	8,034.57	7,437.61	106,897.66	100,273.11
					Hermetic Chillers				
80	72.73	4.08	5.18	16,110.65	15,128.80	866.90	802.52	16,977.55	15,931.32
500	454.55	4.40	5.59	54,000.00	50,708.11	3,825.00	3,540.84	57,825.00	54,248.95
1000	909.09	4.60	5.84	85,324.46	80,122.43	6,706.03	6,207.73	92,030.49	86,330.22
1200	1,818.18	4.40	5.59	134,819.70	126,600.06	11,757.10	10,883.58	146,576.80	137,483.64

[*]System A Capacity is the capacity required for the building. System B is the standard capacity requirement which gives the identical output as System A operating with 60°F condenser water rather than the 80°F water produced by cooling towers.

Performance and capacity variations with temperature taken from Americal Air Filter Catalog AHU-1-121-JAN-02.

Performance (COP) is based on chiller thermal output divided by chiller power input. Values do not include water flow pumping or air flow distribution.

Unit cost data and installation man-hours taken from J. E. Christian, Central Cooling - Comprehensive Chillers, ANL/CES/TE78-2, 1978.

[**]$17 per man-hour is assumed.

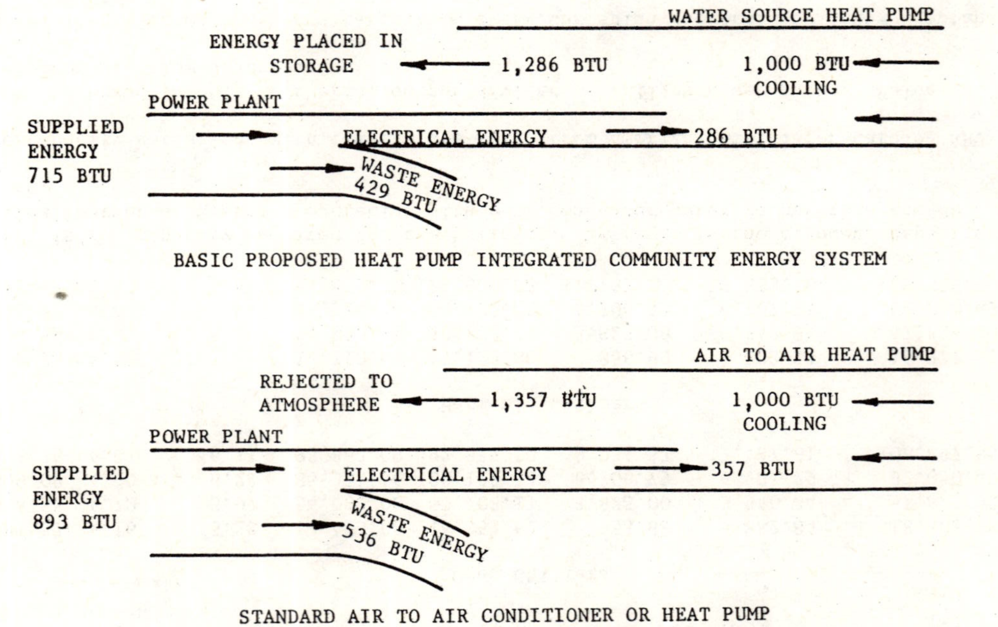

Fig. 9.6. General performance summary of cooling systems.

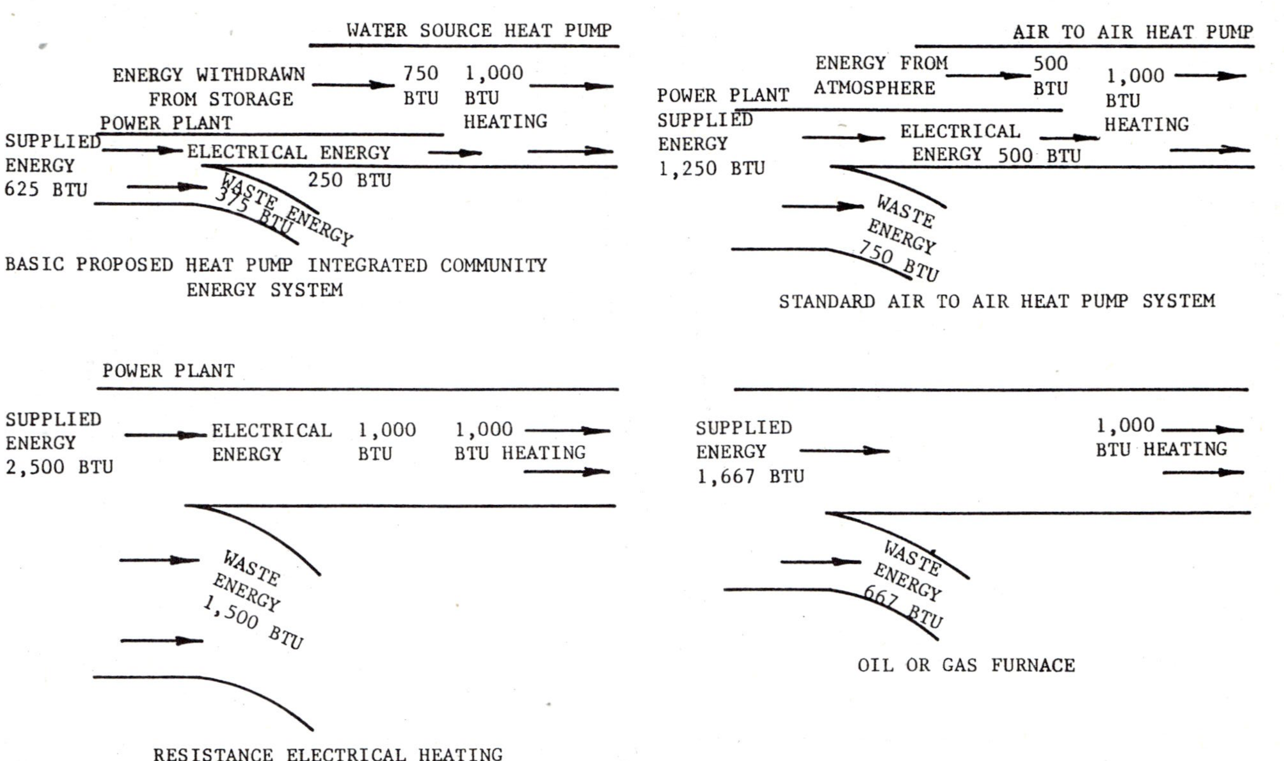

Fig. 9.7. General performance summary of various heating systems.

$$Q_{Heating}/Q_{Cooling} = 2.0$$

This means twice the cooling load is available compared to the heating load. Figure 9.8 gives the ratio of heating load to cooling as a function of the heating and cooling coefficients of performance.

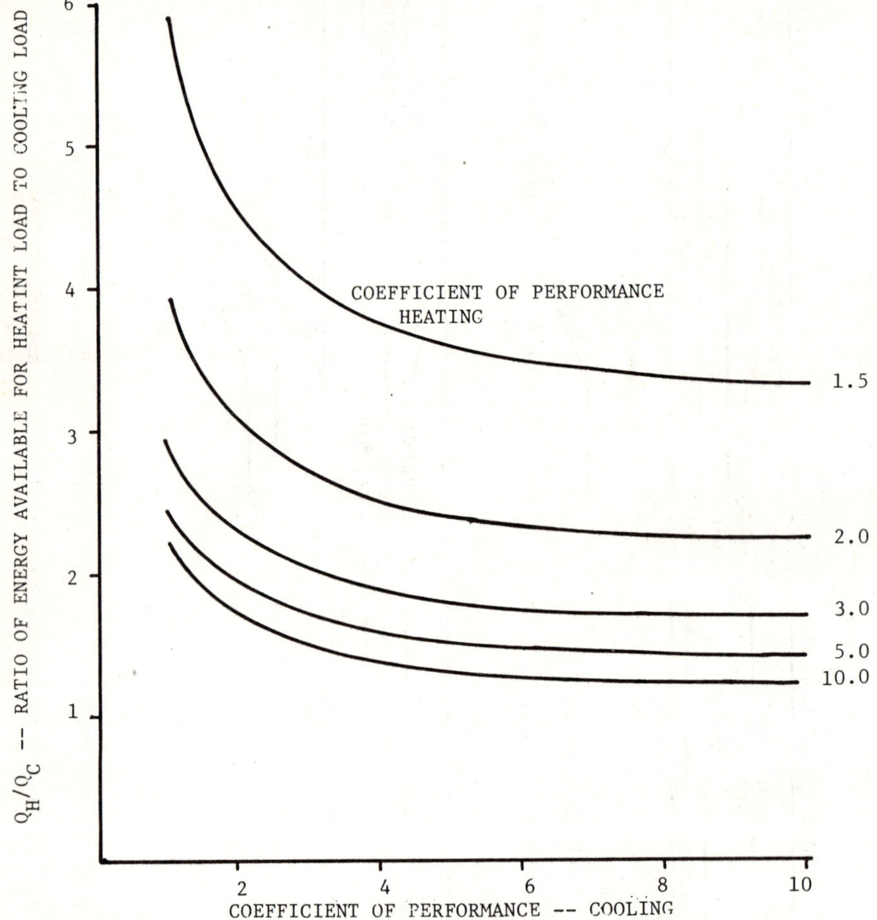

Fig. 9.8. Heating to cooling ratio availability as a function of heat pump performance.

For most parts of the country extra heat must be added to the system for family dwellings. For major buildings with central cores, air conditioning is utilized the year around in most parts of the country. In many northern areas, major buildings can be matched with residential homes in community aquifer systems to balance the loads. In purely residential areas, extra energy must be added to the system. In the South cooling loads dominate and excess energy must be removed from the system.

The basic methods to add energy to heat pump systems requiring supplementary energy are:

1. Waste Heat Sources--This requires placing a heat exchanger between the source and the water distribution system. Any source such as power plant cooling water, waste heat from laundry runoff, etc. can be utilized. In the home, waste

heat from showers, sinks, dishwashers, washing machines, etc. can be channeled through a heat exchanger. This exchange can be on a twelve-month basis even though energy usage is for a shorter period. Energy discarded in the summer can be used in the winter. The temperature can be controlled through a system of thermostatic valves between the two water flows.

2. Solar Energy--Low temperature, low cost solar panels can be used for supplemental energy. Possibilities are hard molded rubber solar panel, a solar panel made from polyethylene bags, rolled rubber panels and shallow-pond collectors. A number of these ponds are pictured schematically in the next chapter. The energy can be collected in the summer months when maximum insolation and maximum collector efficiency are available. Freezing problems will be minimized if collection is limited to the warm months.

3. Matching Buildings in Community Systems--Large buildings have large internal heat loads and lower energy losses per unit area. For example, multi-story buildings have no losses through common ceilings and floors. By matching large buildings with smaller buildings, the heating and cooling loads can be matched with the larger building producing the extra energy required for smaller buildings.

Supplementary cooling is required in many major buildings and across the warmer parts of the United States. Possible methods for removing thermal energy are:

1. Place a cooling tower in the system. Since annual storage exists, a small cooling tower operating continuously is a possibility. The energy removed by a cooling tower must pass through a basic heat exchanger to prevent contamination of the aquifers.

2. Reject a percentage of the hot water after the cooling cycle. This alternative consumes water from the aquifer which is acceptable in some areas. However, a number of systems using water-source heat pumps and then rejecting the water to surface runoff have depleted aquifers.

The heating and cooling loads in an aquifer must be balanced. Otherwise the aquifer will heat or cool continuously. Numerous systems have failed after a few months or years due to energy imbalance. Most of these systems were either 100 percent heating or cooling.

Direct Water-Source Heat Pump Utilization

Water can be pumped directly from aquifers and used for both heating and cooling. Groundwater in the continental United States is in the temperature range for both cooling and heating. In the northern part of the country, the water is sufficiently cold, below 55°F (13°C), so that air conditioning can be provided directly by the water. Figure 2.2 gives the groundwater temperature near the surface for the continental United States. There has been large-scale use of water-source heat pumps. A prime example is Louisville, Ky. (1). The downtown area used water source heat pumps with the water being discharged to surface runoff. After a few years, the 60 ft thick aquifer was depleted of water. With a conversion to conventional heating and cooling systems the water in the aquifer is rising and threatening basements. A major shopping center in Birmingham, Alabama has depleted two aquifers using water source heat pumps with surface discharge of the water. A conversion to air-source heat pumps has been made. In both applications, reinjection of the water could have corrected the problem. However, additional environmental problems would have been encountered.

Direct use of water from aquifers for heat pump systems is feasible if sufficient water is available. Many systems have failed due to depletion of water. These systems generally use over an order of magnitude more water than commonly used for normal domestic consumption, even with lawn watering.

Economics of Heat Pump Systems

Energy economics favor heat pump systems in comparison to all conventional heating and cooling systems. Capital costs are site specific and can be lower or greater than conventional system costs. The additional costs include the well systems and water distribution costs as discussed in Chapter VII. The reduced costs result from:

1. Identical capacity air-source heat pumps have a higher installed cost than water-source units. The water-source units have only one cabinet (no external unit) simplifying installation. Normally in smaller size units no refrigeration connections are required. As a result the water-source units have a lower unit capital cost than conventional heating and cooling systems.

2. With the favorable water-source temperature, smaller capacity units provide the identical output. This not only improves original capital cost but also installation costs.

3. For larger chillers the cooling towers are eliminated.

4. The conventional furnace or backup resistance heat is eliminated or minimized. The units in the proposed systems do not lose capacity with ambient temperatures changes.

After balancing gains and losses, the capital costs for the community heat pump systems can be above or below conventional system capital costs. Each application is site specific and will be different. Chapter XIII defines techniques for combining capital costs and energy costs into a life cycle cost for the system.

References

1. Schaetzle, W.J., Brett, C.E., and Seppanen, M.S., Heat Pump Centered Integrated Community Energy Systems, NTIS No. ANL/CNSV-TM-25 (August 1979).

2. Copsey, M.G., "Analysis of Soil as Heat Source for Heat Pump System," Transactions of ASAE pp. 846-848, (1966.)

3. Grothe, S.P., Sutton, G.E., and Liffler, W.A.," Earth as a Heat Source and Sink for Heat Pump," Engineering Progress at the University of Florida, Vol. IV, No. 6, (June 1950).

4. Krumme, V.W., "Luft-Wasser-Warmepumpe in Einfamilienhaus," Electrowarme International, 32 (January 1974).

5. Michels, V.F., "Probleme Limes Luft-Wasser-Warmepumpe bei tiefen Aussentemperaturm," Bull. SEV/VSE 68, (April 1977).

6. Director of Certified Unitary Air-Conditioners, Air Source Unitary Heat Pumps, Sound Rated Outdoor Unitary Equipment, and Central Systems Humidifiers, and Directory of Certified Applied Air-Conditioning Products, ARI, 1815 North Fort Meyer Drive, Arlington, Virginia 22209, (1980).

7. Personal transmission of Compressor Performance Curves and Tables by Tecumseh, transferred through American Air Filter (1978.)

8. WESCORP Solar Systems Catalog, including Typical Specification Sheets WESCORP Inc. 1979-019 and 1978-098, includes numerous specifications and configurations, from WESCORP, Inc., 15 Sterons St., Andover, Ma., 01810.

9. Christian, J.E., Unitary Water to Air Heat Pumps, ANL/CES/TE/77-10, Oak Ridge National Laboratory , Oak Ridge, Tenn. (July 1977).

10. Christian, J.E., _Unitary Water-To-Air Heat Pumps_, ANL/CES/TE/77-9, Oak Ridge National Laboratory, Oak Ridge, Tenn. (October 1977).

11. _Rising Ground-Water Level in Downtown Louisville, Kentucky, 1972-1977_, USGS (Sept. 1977).

CHAPTER X

SOLAR APPLICATIONS

In general, most solar systems collect thermal energy near the time of usage and energy is stored for a maximum of a few days. The thermal energy storage capacity is limited by volume and capital cost, thereby limiting storage capacity to a few days or a few hours. Storage systems discussed herein use large thermal energy capacities provided by nature in the form of groundwater aquifers. Aquifers can provide large thermal energy capacities at capital costs less than those of the current short-term solar thermal energy storage systems. As a result, storage periods consist of a major annual cycle with minor short term cycles.

The annual storage and large storage capacity can provide numerous solar energy collection advantages. These include:
1. Collection during summer periods when insolation is at a maximum,
2. Collection during summer periods when cloud interference is minimized in certain parts of the country,
3. Collection during summer periods when ambient temperature is a maximum, thereby increasing collector panel efficiency,
4. Collection at lower temperatures since the storage system does not require the packing of energy,
5. Minimizing or eliminating the requirement for a backup heating system,
6. Backup systems can be smaller as they will not be required during mid-winter peak loads but rather during smaller spring loads, and
7. Power company peak loads during the heating season are decreased.

In general, using annual storage for heating will result in solar panel collection area reduction from 50 percent to over 75 percent for various geographical areas in the continental USA. As with heat pump systems, solar energy blends with annual aquifer thermal energy storage into the community energy system.

Figure 10.1 shows a schematic of a solar energy system in parallel with an aquifer thermal energy storge system. When solar energy is available, the energy can be used directly for heating, hot water, etc. The excess energy is stored in the aquifer. When solar energy is not available, energy can be withdrawn from the aquifer.

As with the heat pump systems, an aquifer storage system is applicable in the case of a large building with a major load or a community system which comprises a major load to be technically feasible. The water-distribution system would be a two-pipe distribution system between the buildings, collectors, and storage. Since the temperatures will be well above natural ground temperature, the

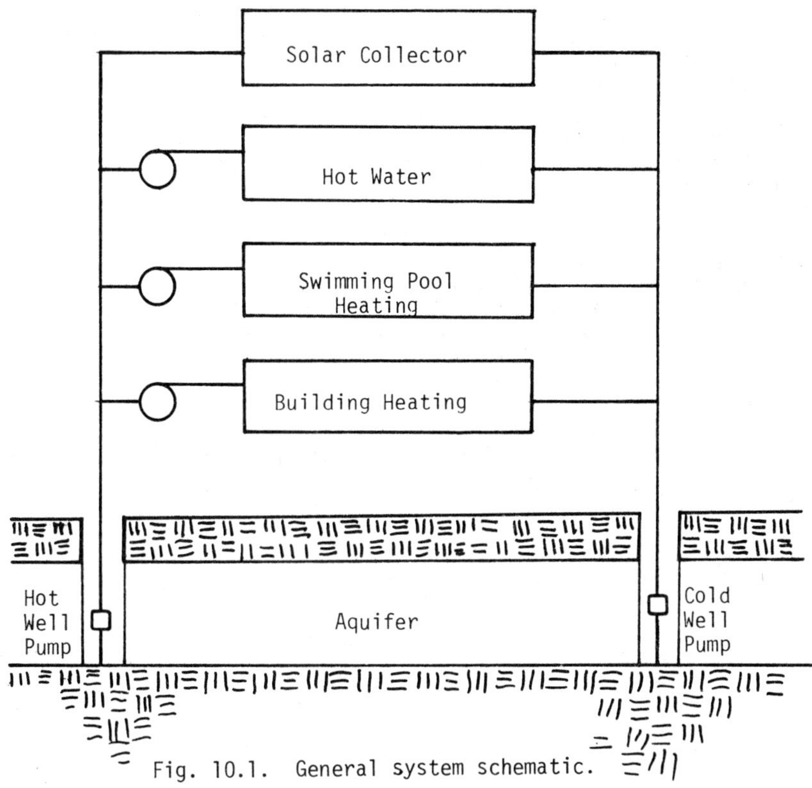

Fig. 10.1. General system schematic.

pipelines must be insulated. Smaller pipes, up to two inches in diameters, normally will be covered with one-half inch thick foam rubber sleeves. For the primary distribution system, up to eight-inch diameter pipes, two inches of fiberglass insulation is normal. In some cases, one inch and two inches of calcium silicate insulation is being used for ground compatibility. As temperatures increase, above 150°F, insulation thickness should increase. Not only is the loss of energy important but also the degrading of energy. The cost data for using PVC pipe for the distribution system are provided in Fig. 7.3. The water distribution system can be hydraulically controlled using pressure sensors for pump on/off sequences with pressurized storage. A second alternative is the use of variable-speed pumps with computer control to satisfy supply and demand.

In community systems the solar panels can be placed on a few homes rather than one panel per individual home. A central panel system is an alternative. This is especially feasible with annual collection as collector size decreases. Massed panels minimize installation costs and the requirement for more controls. This in turn increases reliability and minimizes maintenance.

A number of independent community systems with collectors and aquifer thermal energy storage can be connected in parallel. This duplication increases overall system reliability. If the pumps in one aquifer fail the attached systems provide backup. This eliminates the need to change a pump at 2:00 A.M. in -40°F (-40°C) weather. Most maintenance could be handled on a routine (preventive) maintenance schedule. A community energy system has many advantages with the large energy storage capacity available in aquifers.

Solar Thermal Energy Collection

Design of collection systems for solar energy vary as a function of collection temperature. Temperatures above 500°F (250°C) can be stored in aquifers. This high temperature storage technique is discussed in Chapter XII. To collect energy in this temperature range, concentrating collectors are required. One-dimensional parabolic collectors are normally utilized. For temperatures near and below the boiling point of water, flat plate collectors are normally employed although some focusing collectors have been utilized. For the higher temperatures in this range and for cold weather collection flat plate collectors normally have one or more layers of glazing, selective coatings on the surface, and good insulation. For summer collection of moderate temperatures (100°F to 140°F) (38°C to 66°C) unglazed swimming pool-type collectors are utilized. An analysis of high temperature collection and moderate temperature collection is included. The effects of annual collection are emphasized. Lower temperature swimming pool-type collectors are also reviewed.

High temperature energy can be used for process heat or to produce low temperature steam. Numerous industrial applications exist in this area. It is the area of maximum energy usage in the industrial sector. An analysis of aquifer storage specifications for this temperature range is made in the next chapter.

Moderate-temperature energy can be used for hot water, heating, air conditioning, etc. Collection and use of this energy is normally on a daily basis.

Low-temperature collection can be used for hot water heating, swimming pools, etc.; however, the energy is difficult to store in normal systems due to system size. To make the storage cost-effective, the energy must be packed with the maximum possible storage temperature difference. The aquifer thermal energy storage system eliminates this problem as nature provides a large storage system at economical cost. It is feasible to store water with temperatures as low as 60°F (16°C). This temperature range could be used for a solar-assisted heat pump system. Summer thermal energy collection using swimming pool-type collectors for temperatures in the range of 90°F to 140°F (32°C to 60°C) with aquifer storage on an annual basis is a definite potential.

High Temperature Collection

Focusing collectors are required for high temperature (over 212°F) (100°C) energy collection. A cylindrical parabolic collector is normally used for the temperature range of 212°F to 500°F, (100°F to 260°C). The upper temperature limits are aquifers storage temperature limits. A parabolic collector sketch is shown in Fig. 10.2. The following data and description are for a Honeywell (1) parabolic solar collector.

The collector consists of two basic components: the concentrator and the receiver. The concentrator is made of lightweight acrylic and has the shape of a half-section of a parabola. The acrylic has demonstrated long-time durability to solar exposure. The flat profile of half a parabola minimizes the aerodynamic drag, especially in the horizontal stored plane.

The receiver is an absorber tube mounted in an evacuated glass tube. The absorber tube has insulation over 3/4 of its circumference. Solar flux concentrates on the other 1/4 of the tube through the outer glass tube. The absorber tube is coated with a selective coating of black chrome over bright nickel. The selective coating gives an absorbtivity of 0.94 and an emissivity of 0.12. The collector axis can be mounted parallel to either a north-south line or an east-west line. For a north-south line axis mounting the collector moves through a complete cycle going from east to west each day. The axis of rotation throughout the United States is normally inclined with respect to a horizontal plane to increase energy input. An east-west collector axis mounting requires a

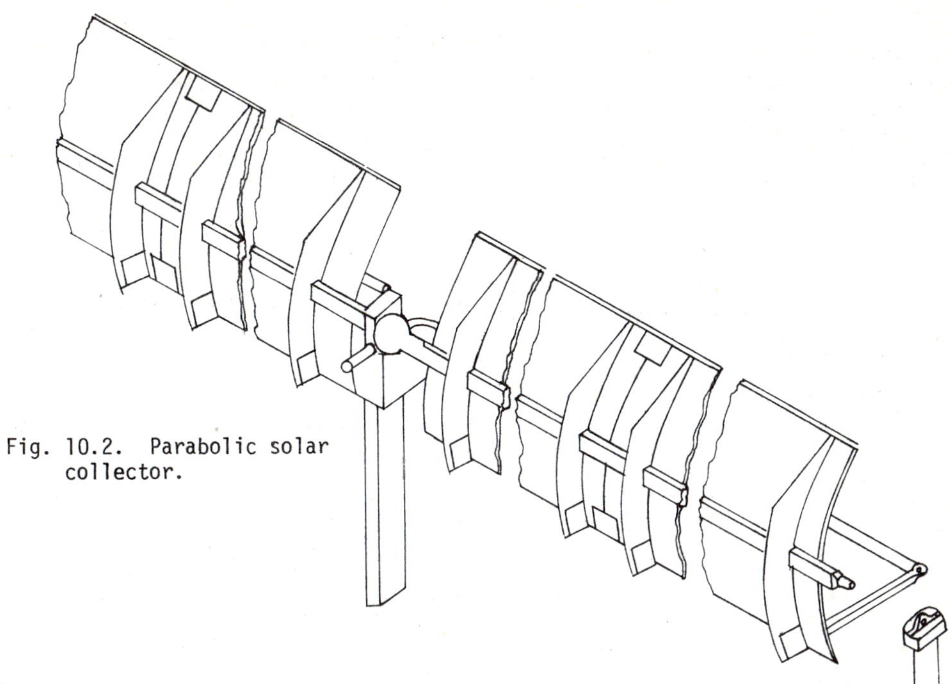

Fig. 10.2. Parabolic solar collector.

complete cycle only once a year. Small changes in orientation are required only every few days. The axis of rotation for east-west orientation is normally mounted in a horizontal plane. An inclined north-south mounting rather than an east-west mounting is normally used to maximize energy absorbed, however the control system is more complex.

The concentrating collector has a slightly lower base collection efficiency than the flat plate collector but is not effected by ambient temperatures. An efficiency curve is shown in Fig. 10.3. A number of constant losses such as reflectance, transmittance, and absorption are shown in the figure. The prime efficiency variation is with tube inlet temperature. The dropoff in collector efficiency with inlet temperature is very small in comparison with the flat plate collector.

Calculations have been made using a modified form of the F chart technique of Beckman (2). Latitude and empirical insolation data from Tuscon, Arizona, and the Honeywell efficiency curve in Fig. 10.3 are used for input. The collectors are oriented on a north-south axis and inclined at $0°$, $10°$, $20°$ and $30°$ from the horizontal. For these calculations the incident radiation is isolated from total

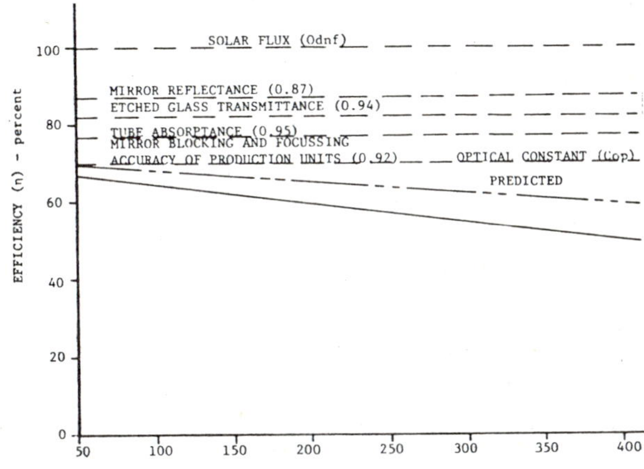

Fig. 10.3. Performance of the concentrating collector.

radiation and used for the insolation value. The average daily collections by month were calculated and are plotted in Fig. 10.4 along with the annual averages.

The curves show the variation of energy collected throughout the year. Using annual thermal energy storage requires less collector area for a constant demand as energy can be stored for off-season demand. The average annual daily energy collection is noted on the figure.

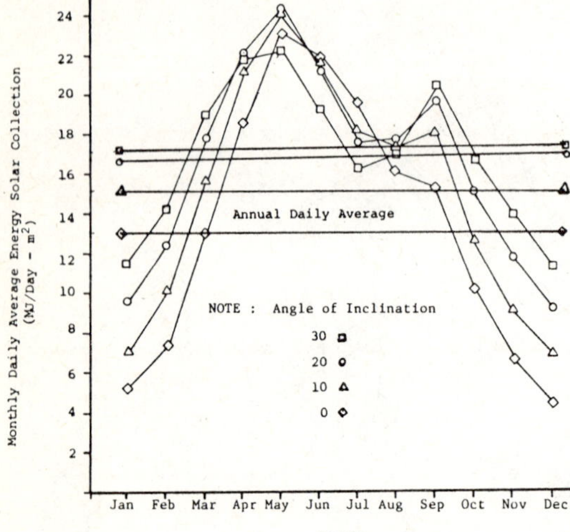

Fig. 10.4.   Solar energy collection for Tucson, Arizona.

An example of an application is the solar power plant system shown in Fig. 10.5 (3,4).   The aquifer provides storage on:

1.   A daily basis for night-time energy requirements,

2.   A few-day basis for bad weather energy requirements, and

3.   An annual basis for lower winter insolation requirements.

It is recommended in this case to begin storage as soon as some collectors are in place to condition the aquifer. Conditioning the aquifer means heating the storage volume before actual storage begins. This provides a higher operational storage efficiency, for operational and energy storage and an energy reserve for unexpected emergencies.   An analysis of the problems involved in high-temperature storage is discussed in Chapter XII.

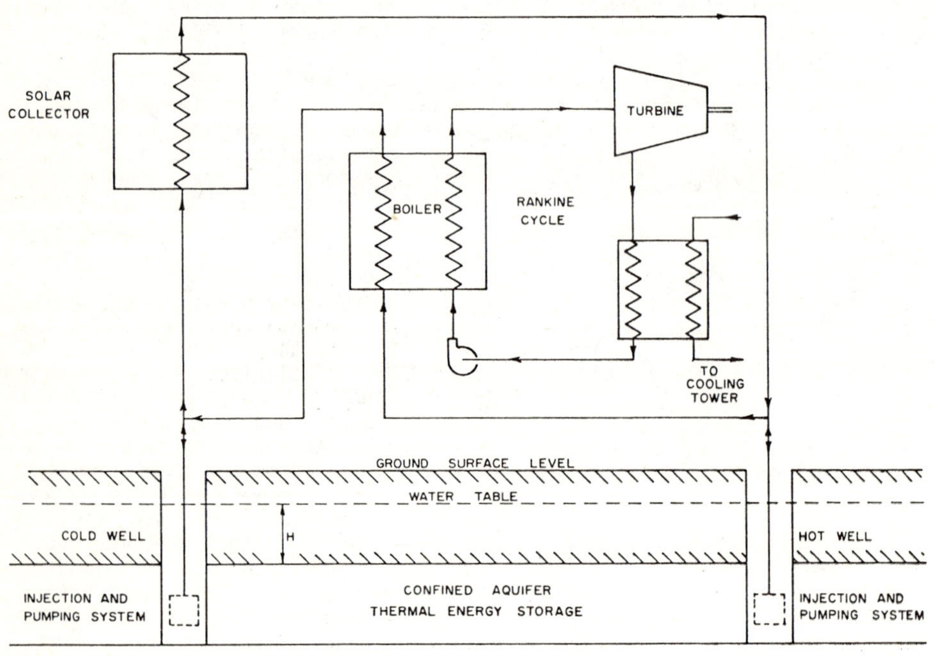

Fig. 10.5.   Solar power plant system – schematic.

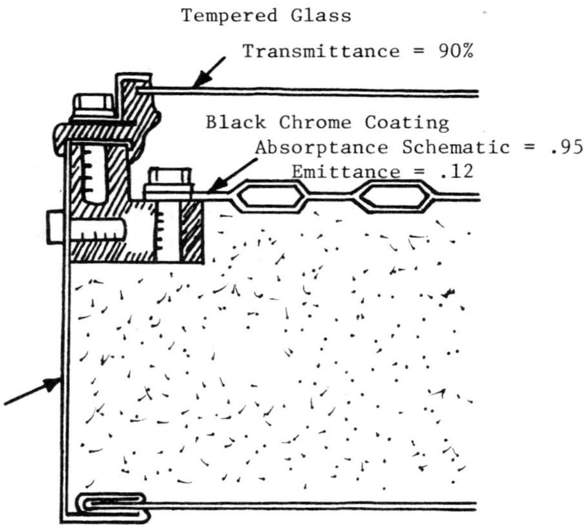

1. Insulating Mounting Block

2. Fiberglass Insulation

3. Galvanized Steel Collector Box

Characteristics	Collector
Manufacturer	Chamberlain Mfg. Co.
Gross dimensions (m)	2.14 x 0.92
Gross area (m^2)	1.96
Effective aperture area (m^2)	1.79
Weight per gross area (kg/m^2)	37.0
Pressure drop across collector (Pa)(at operating mass flow rate)	1.6
Cover plate assembly	
Number of glass plates	1
Material	Fourco-Cleartemp
Dimensions (m)	2.08 x 0.86
Solar transmittance (single glass)	0.90
Hmeispherical emittance	0.88
Coefficient of expansions (mm/mm/°C)	2.78×10^{-6}
Specific heat J/(kg.°C)	Unknown
Absorber Plate	
Material	Mild steel
Dimensions (m)	2.09 x 0.86
Flow configuration	19 parallel pass
Coating	Black chrome
Solar absorptance	0.94
Hemispherical emittance	0.12
Between cover and absorber(mm)	19.0
Insulation	
Material	Glass-fiber
Density (k g/m^3)	80
Thickness (mm)	76.2

Fig. 10.6. Pertinent collector characteristics (from the manufacturer's literature).

An economic analysis (4) indicates the cost of the aquifer thermal energy storage system provided by nature is only 0.4 percent of the capital required for the total solar power plant. The storage capital cost including wells, well casing, and pumps and plumbing for a 40 MW plant come to $1,600,000. This is a fraction of a conventional night-time energy storage system.

Medium Temperature Collection

In the medium temperature collection regime, flat plate collectors are very common. These collectors are simply mounted stationary and absorb incident reflected and diffused radiation. They even operate during periods when the sun is blocked by clouds but at a lower level of efficiency. This is important in many geographical locations. Their effective upper temperature limit is around the boiling point of water, although many are used to collect energy at over 250°F (120°C). In general these are the widely used collectors for heating buildings and providing hot water. When air conditioning is included, focusing collectors are often utilized.

A typical single glaze liquid flat plate collector evaluated for the DOE by Streed et al (5) with manufacturer's specifications is given in Fig. 10.6. The collector was choosen by the authors for being a commercially available collector from an established manufacturer employing established manufacturing and quality assurance techniques. The collector components are the collector box, the collector plate, and the glazing above the collector plate. Normally one or two layers of glazing cover the collection plate to provide a greenhouse effect and minimize convection from the system. The back of the plate normally contains two to four inches of insulation to minimize energy losses through the back. An insulated connector to the collector box is a critical item. The collector plate contains passageways to transfer the collected energy. The plate which transfers the energy to the collection liquid must be a good condutor. The selective coating should maximize absorption and minimize emissivity. The collector in Fig. 10.6 has an absorptivity of 0.94 and an emissivity of 0.12.

The efficiency of these collectors is sensitive to ambient temperature, collection temperature, and magnitude of insolation. The efficiency curve for the collector in Fig. 10.6 is given in Fig. 10.7. The dependent variable is a ratio of the temperatures and insolation. The optimum inclination angles for the collector are given in Fig. 10.8. Variations are given to optimize summer, winter, or annual collection.

A modified F chart method of predicting solar energy collected is by Lui and Jordan (6). The actual data in the paper for a horizontal surface are based on a five-year average of collected data by NOAA. The system uses a combination of direct, diffuse, and reflected radiation to predict the values for an inclined surface. The hourly peak energy collection is per Iqbal (7). The maximum average daily rate was utilized for insolation on the collector efficiency curve. This is rationalized on the basis that during many days the incident radiation is not sufficient for energy collection. In turn, on days of sufficient incident insolation, the peak rate of insolation is appreciably

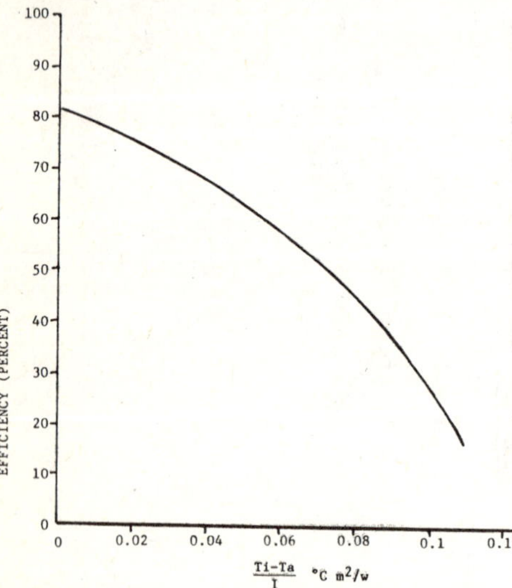

Fig. 10.7. Collector efficiency curve.

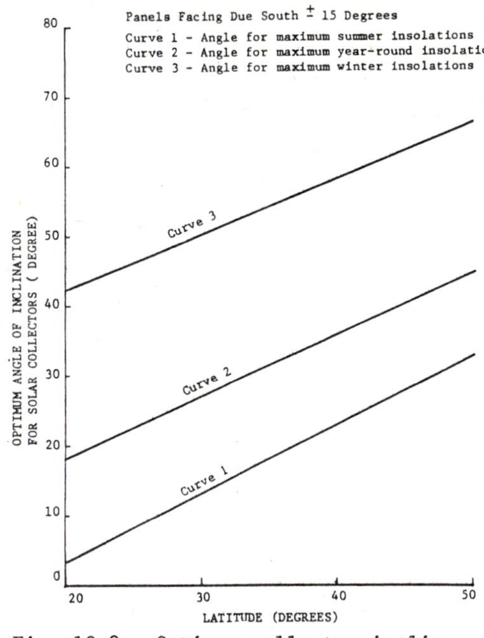

Fig. 10.8. Optimum collector inclination vs. latitudes.

larger than the monthly maximum average value. As a result, the peak average is used as the average incident insolation during periods of energy collection. The ambient temperature used for collector efficiency is 0.707 times the value between the daily average temperature and daily average maximum temperature listed in Reference 8 and based on the 1941 to 1970 period. A computer program for making the calculations is available.

Table 10.1 gives the summer, winter, and annual collection values for optimum annual inclination of the collector. In general, the summer collection is over two times the winter collection. With the capability to collect and store this energy on an annual basis, collector area can be reduced by over two-thirds in most heating applications. Since collector cost and installation is the major capital cost, this should also be reduced by two-thirds. In addition, with the large storage system, no backup system is required and the aquifer storage system cost is less than the normal overnight storage system cost.

An economic analysis is made for systems in Birmingham and Boston to estimate economic feasibility. The aquifer thermal energy storage system and water distribution system are assumed to cost an equal amount, $2,387 per home, in both cities (community system from Chapter VII) and water pumping cost are assumed to be $30/yr. as derived previously. The model homes are assumed to use 10^8 Btu/yr in each city allowing for different insulation levels. The value also allows for 80 percent recovery in storage. The analysis is given in Table 10.2. Straight-line depreciation over twenty years is assumed. Electricity cost is estimated at $0.05/kWhr and oil cost is estimated at $4.50/million Btu ($.50/gal).

The solar annual storage system is economical when compared to resistance heating; however, oil heating, gas heating or a heat pump are still more economical at the utilized fuel prices. With rising energy costs and a more economical collector ($8/ft^2 installed), the solar system becomes economical. The collector cost comparison with a conventional solar system collecting energy only during the winter months in Table 10.2 demonstrates the economic advantage of the annual energy storage system. Remember, the short term storage system costs more per home than the proposed annual storage system.

The large annual storage allows twelve-month energy collection which appreciably reduces required collector area and eliminates the backup system. The extension to twelve months allows collection at maximum insolation and at maximum collector efficiency. The cost of the proposed solar collection system with annual storage is appreciably less than that of a conventional solar system for heating.

Low Temperature Collection

Low temperature collection refers to temperatures under 140°F (60°C). With annual thermal storage and the relatively low collection temperatures, basic swimming pool-type solar collectors are a possibility. (Summer usage eliminates the freezing problem and the requirement for glazing thereby allowing direct

TABLE 10.1. Annual Solar Energy Collected for Twenty American Cities

Average collector temperature: 125°F

Location	Latitude, deg	Optimum tilted angle, deg	Summer energy collected Btu/ft^2	Winter energy collected Btu/ft^2	Annual total energy Btu/ft^2
Birmingham, AL	33.3	30.2	233430	119560	352990
Tucson, AZ	32.1	29.1	309290	232300	541590
Los Angeles, CA	33.6	30.4	252080	178300	430390
Denver, CO	39.4	35.5	263830	179070	442900
Miami, FL	25.5	23.0	233480	206050	439530
Atlanta, GA	33.4	30.2	221870	123790	345660
Chicago, IL	41.6	37.4	203970	54270	258240
New Orleans, LA	29.6	26.8	181770	117430	299200
Boston, MA	42.2	37.9	180500	36620	217120
Detroit, MI	42.1	37.8	206700	42610	249310
Saint Louis, MO	38.4	34.6	223200	79870	303070
New York, NY	40.5	36.4	169620	36310	205930
Tulsa, OK	36.1	32.6	230690	114230	344920
Portland, OR	45.4	40.6	183250	27120	210370
Philadelphia, PA	39.5	35.6	207330	73950	281280
Nashville, TN	36.1	32.6	224080	80020	304100
Houston, TX	29.6	26.8	238800	143370	382170
Salt Lake, UT	40.5	36.4	249790	36360	336150
Seattle, WA	47.3	42.1	200980	29930	230910
Milwaukee, WI	42.6	38.2	208170	45150	253320

TABLE 10.2. Economic Analysis for Proposed System

	Boston	Birmingham
Energy usage	10^8 Btu/yr.	10^8 Btu/yr.
Collector area required	440 ft^2	280 ft^2
Installed collector cost ($15/$ft^2$)	$6600	$4200
Storage and distribution system cost	$2850	$2850
Total solar capital cost*	$9450	$7050
Annual capital cost (10% interest plus 5% depreciation)	$1418	$1058
Annual pumping cost	$30	$30
Total annual solar cost	$1448	$1088
Annual energy cost (resistance heating)	$1465	$1465
Annual oil heating	$750	$750
Collector area required (without annual storage)	1337 ft^2	740 ft^2
Installed collector cost ($15/$ft^2$)	$20,050	$11,100

*Maintaining costs and air circulation costs are assumed identical in both areas. The furnace capital costs could be deducted from solar capital costs or added to electric or oil heating costs.

system water circulation). Shallow solar ponds are also effective in this temperature range. Figures 10.9 and 10.10 (9, 10) show sketchs of swimming pool-type collectors. The cost of these usually averages around $3/ft^2 with an additional $2/ft^2 installation charge when mounted directly on a roof for summer use only.

Rolls of vinyl rubber collectors with internally molded tubes are in the same price range. These collectors are constructed so a layer of glazing can be utilized for slightly higher temperatures and extended collection. Freezing will not damage many of these collectors.

Figures 10.11 gives a possible configuration for a solar pond collector along with relative price estimates. References 11 through 14 discuss this solar collection concept. Note that some industrial estimates of solar pond collector costs have been as high as $20/ft^2.

The large aquifer thermal energy system makes the collection of solar energy during the summer season available for annual heating utilization. This also allows the use of more economical collectors with less complicated controls as freezing is not a problem. Solar assisted heat pumps are an ideal application of this low temperature energy.

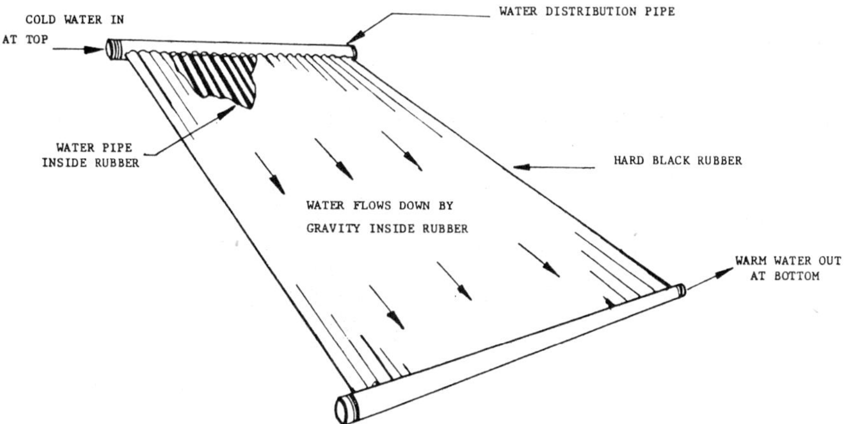

Fig. 10.9. Sketch of hard rubber molded solar panel without glazing.

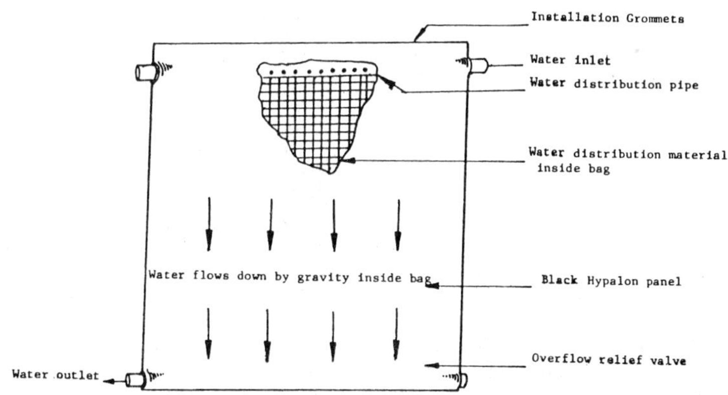

Fig. 10.10. Panel configuration for solar bag collector.

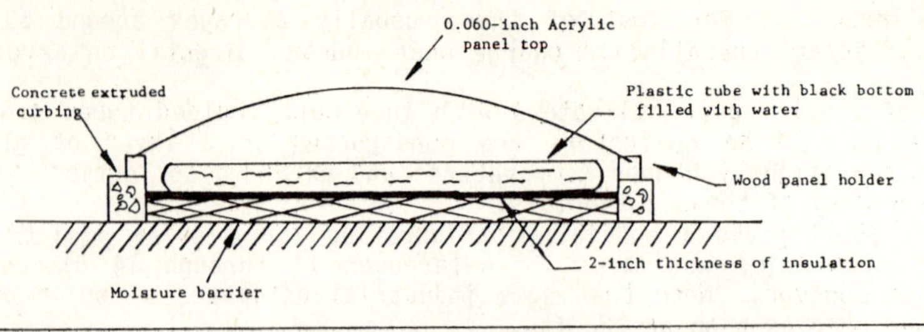

Solar Pond Costs - ERDA/Sohio Project.

	Low cost - Air Inflated (PVC)	Long life - Panel top (Acrylic)
1. Site preparation	$0.10 per ft.^{2}	$0.10 per ft.^{2}
2. Moisture barrier	0.04	0.04
3. Bottom insulation	0.37	0.37
4. Water bag and glazing assembly	0.70	0.77
5. Curbing and mounting		0.33
6. Assembly hardware and installation	0.08	0.05
7. Piping and fittings	0.05	0.05
8. Pumps and motors	0.06	0.06
9. Reservoirs - hot	0.22	0.22
10. Reservoirs - cold	0.20	0.20
11. Instruments and controls	0.05	0.05
Total	$2.87 per ft.^{2}	2.24 per ft.^{2}

Fig. 10.11. Typical solar pond installation with relative costs.

Summary

Large aquifer thermal energy storage is applicable over a wide range of potential solar energy collection temperatures. Annual storage allows collections at the maximum insolation and collector efficiency conditions. This storage system has tremendous potential to advance the solar energy field.

References

1. Gupta, B.P., *Development and Evaluation of a Medium Temperature Concentrating Collector*, Energy Resource Center, Honeywell, Inc., Minneapolis, Minn., (1977).

2. Bechman, W.A., Klein, S.A., and Duffie, J.A., *Solar Heat Design*, Wiley, New York, (1977).

3. Ansari, J.M., *High Temperature Thermal Energy Storage in Aquifers with a Solar Power Plant Application*, Ph.D. Dissertation, University of Alabama (1980).

4. Schaetzle, W.J., Brett, C.E., and Ansari, J.M.," Thermal Energy Storage in Aquifers for a Solar Power Plant, " *Proceedings of 2nd International Conference on Alternative Energy Sources*, University of Miami, (Dec. 1979).

5. Streed, E.R., Hill, J.E., Thomas, W.C., and Dawson, A.D., "Results and Analysis of a Round Robin Test Program for Liquid-Heating Flat-Plate Collectors," *Solar Energy*, Vol. 22, No. 3, (1978).

6. Liu, Y.H., and Jordan, R.C., "A Rational Procedure for Predicting the Long-Term Average Performance of Flat-Plate Solar Energy Collectors," _Solar Energy_, Vol. 7, No. 2, (1963).

7. Iqbal, M., "Hourly vs Daily Method of Computing Insolation on Inclined Surfaces," _Solar Energy_, Vol. 21, No. 6, (1978).

8. _Local Climatological Data_, NOAA, National Climatic Center, Ashville, N.C., (1977).

9. _Burke Solar Heater_, Burke Rubber Company, San Jose, California (1978).

10. _Energy From the Sun in Your Back Yard_, Fayco, Inc., Menlo Park, California (1977).

11. Cay, J.A., Clarke, A.F., Dickenson, W.C., and Iontuono, A., _Industrial Process Heat from Solar Energy_, IECEC75 Record (1975).

12. Labor, II., "Solar Ponds: Large-Area Solar Collectors for Power Production," _Solar Energy_, Vol. 7, No. 4, pp. 189-194 (1963).

13. Weinberger, H., "The Physics of the Solar Pond," _Solar Energy_ Vol. 8, No. 2, pp. 45-56 (1964).

14. Dickinson, W.C., Clark, A.F., Day, J.A., and Wouters, L.F., "The Shallow Solar Pond Energy Conversion System," _Solar Energy_, Vol. 18, pp. 3-10 (1976).

CHAPTER XI

DIRECT COOLING

The direct cooling system has the greatest potential for direct energy saving utilizing aquifer thermal energy storage. In the majority of direct cooling applications, there is also a major savings in capital cost. The system is known as the "free cooling system." During cold-weather periods water is chilled in cooling towers (wet or dry) or cooling ponds to provide summer air conditioning. Temperatures near the freezing point of water can be utilized. Chilled water is stored in the aquifers on an annual basis and retrieved as required for air conditioning. The system eliminates the chiller (refrigeration cycle) completely. The major capital cost and energy consumption component are eliminated. A water distribution system and a pair of wells replace the chiller. In most cases the system capital cost shows a large percentage decrease. In many cases the energy used for chilling is cut by over 90 percent. The system availability is site specific, a suitable aquifer must be available.

System Description

The system is a two-pipe system connecting the cooling coils, cooling tower, and well pairs. Figure 11.1 is a schematic of the system. The basic water temperature range is 40°F (4°C) for the cold pipe and 60°F (15°C) for the warm pipe. Since the pipes are near ground temperature, insulation is normally not required for either pipe when buried. Condensation will form on both pipes when exposed to warm humid air, such as in an underground tunnel or the interior of a building. Insulation is required in these cases.
The two-pipe system could be half of a four-pipe water distribution system, where two pipes are for heating (solar or district heating for example) and two pipes are for the "free cooling" system. The two hot pipes must be insulated under these conditions.
The water distribution system is similar to the system for heat pumps as shown in Fig. 9.1 (heat exchange with cold water can be for a small individual unit in a home, a large central unit in a major building or a distributed system in a major building). The chiller (air conditioning cycle) is replaced by a water-to-air or a water-to-water heat exchanger. A standard A-frame type unit for water-to-air units is sufficient. For a water-to-water unit, a counterflow tube-in-tube heat exchanger gives maximum efficiency. With the A-frame type heat exchanger, a degree of counterflow can be achieved by linking the two halves in series. Since the water heat exchanger (even forced convection) has a lower

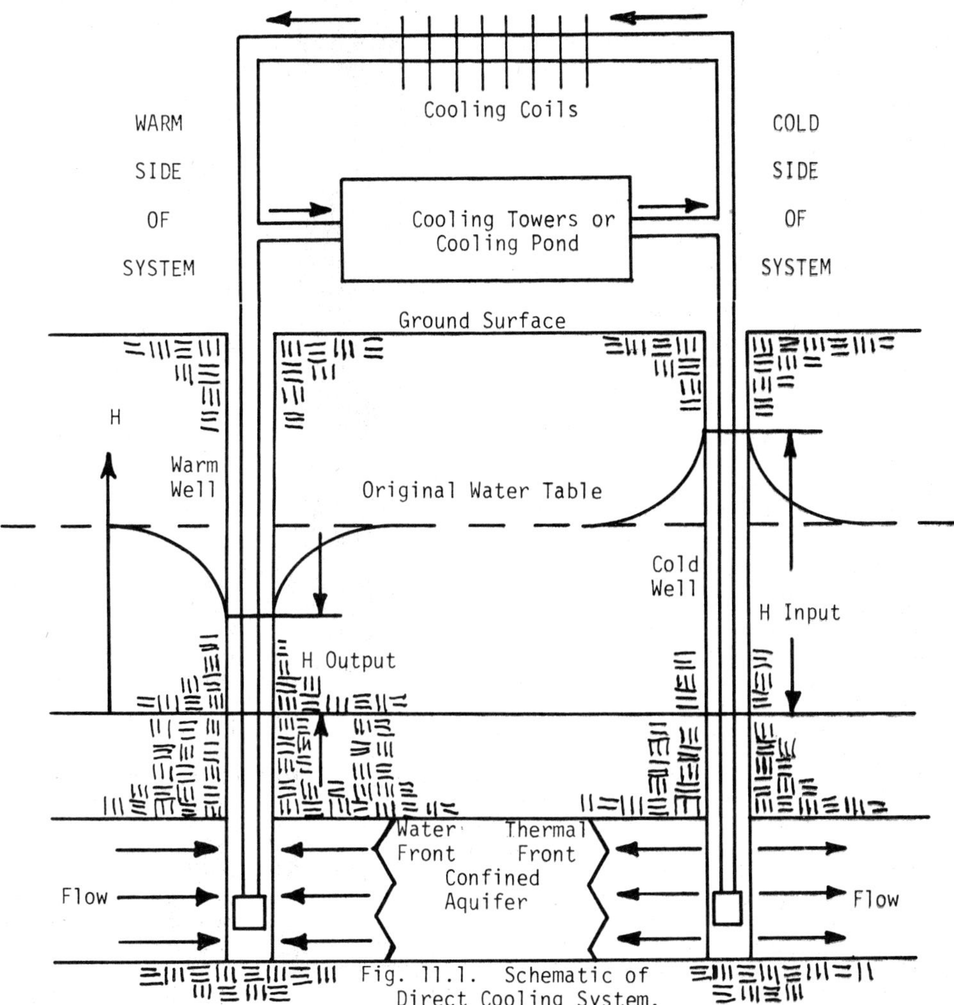

Fig. 11.1. Schematic of Direct Cooling System.

coefficient of heat transfer than boiling or condensation, an increased capacity (more heat exchange area) heat exchanger is required.

"Free cooling" where water from cooling towers is directly utilized during periods when the wet bulb temperature is below 55°F (13°C) has been incorporated in numerous systems. Water bypasses the chiller and is used directly in the building to cool heat exchangers. A number of buildings in Atlanta, Georgia, and the Dallas/Fort Worth Airport use this energy-saving technique. An energy payback of $60,000 in the first three months, on a $20,000 investment, was realized at the Dallas/Fort Worth Airport. In addition, a number of environmental problems, namely gas fixed plume suppression requirements, were eliminated. A number of problems occurred including: cooling tower gearbox lubrication, silt sedimentation, structural failure due to freezing, fouling, and corrosion. The various chemicals utilized to control cooling tower water quality cannot be used where water is injected into aquifers. The phosphonate base chemical treatment for protection from corrosion and fouling is not acceptable to the EPA (1) for injection into groundwater (a heat exchanger between the cooling tower and distribution water and aquifer water is a possible solution).

142

For a 20°F (9°C) temperature difference a flow rate of 10 lb/min (4.5 kg/min) or 1.2 gpm of water is required for one ton of air conditioning. It should be noted that 20 to 80 percent of this energy is required for humidity control.

It is advisable to use outside air for air conditioning when temperatures reach 55°F (13°C) or lower. This also represents "free cooling" as zero energy is required for chilling.

A secondary application can be refrigeration to below 40°F (4°) for normal refrigeration. In this case the water is used to cool the condenser side of the refrigerator's cycle. The cooler water allows the units to operate at higher performance levels, removes the rejected heat directly from the building without additional air conditioning loads, and allows lower-capacity units. As water input temperature decreases the unit performance and capacity increase. The performance and capacity changes as a function of temperature are similar to those presented by Figs. 9.4 and 9.5 for heat pumps. The capacity change allows the use of smaller-capacity units, decreasing capital cost in addition to energy cost savings.

The recommended aquifer installation is the paired well system, allowing control of the cold-front location between wells. A major advantage is storage in the second well of warm water at 60°F (15°C) after the 40°F (4°C) has been utilized for chilling. Thereby the chilling process reduces water temperature from 60°F (15°C) to 40°F (4°C) rather than reducing higher water temperature to 40°F (4°C).

In the cooling process water is cooled in either a cooling tower or cooling pond during weather when the wet-bulb temperature is below the desired temperature. Table 11.1 gives the average number of annual hours below specific wet-bulb temperatures for a cross-section of cities throughout the United States. Note that there are not sufficient hours for cooling in Miami, Florida, or Los Angeles, California. Other than parts of Florida and California, sufficient cooling hours are available, but are limited in southern parts of the United States.

Most towers are designed for constant flow rates in relatively warm weather. Cold operation is limited: heaters are placed in sumps to prevent freezing. Reference 2 notes the approach temperature to wet-bulb temperature increases at lower temperatures. As temperature decreases, the ability of air to hold water vapor decreases. For example, at 80°F (27°C) air holds four times as much water vapor as at 40°F (4°C). This means a greater percentage of the cooling is sensible cooling rather than evaporative cooling for low-temperature water. For the identical quantities of cooling a larger cooling tower is required at lower temperatures. However, in this system the cooling tower operates as a function of cold weather hours. The tower does not have to be designed to meet the peak cooling load, only the average load. The total cooling tower heat system is decreased since the chilling cycle compressor energy input is eliminated, not dissipated, by the cooling tower.

With cold-water storage, the cooling tower water flow rate is not restrained to a constant flow volume. As soon as wet-bulb temperature moves below the required temperature, water is cooled at the minimum flow rate. As the temperature drops, water flow rates are increased to take advantage of the added cooling capacity. As temperature decreases, sensible cooling increases relative to evaporative cooling which also increases. At sufficiently low temperatures after maximum water rate has been reached, the air flow can be throttled by inlet vanes to prevent excessively low temperatures and freezing. This automatically lowers the blower power. Since cooling operation is a function of temperature only and not of cooling load, freezing should be a minor problem during steady-state operation. Precautionary measures must be taken for any variation in operation. Presently, electric heaters are installed in cooling tower basins to ensure freezing will not occur. Pipe freezing, entire towers enclosed in ice, and major structural damage have occurred in cooling towers which cool water to only 50°F (10°C). The problem is much more critical as water is cooled to lower temperatures.

TABLE 10.1. Annual Hours Below the Given Temperature*

City, State	Wet bulb temperature °F (°C)									
	60(15.5)	55(12.7)	50(10.0)	45(7.2)	40(4.4)	35(1.7)	30(−1.1)	25(−3.9)	20(−6.7)	15(−9.4)
Birmingham, AL	4799	4019	3346	2577	1675	1028	564	237	83	20
Mobile, AL	3383	2688	2085	1502	792	416	189	72	8	2
Tucson, AZ	5563	5834	4936	3590	2304	1618	1297	158	2	0
Los Angeles, CA	6395	4074	2114	972	192	25	3	0	0	0
Denver, CO	7676	6879	6069	5102	4065	3128	2257	1457	711	289
Miami, FL	1179	623	351	165	30	3	0	0	0	0
Atlanta, GA	4616	3819	3114	2412	1410	944	485	229	71	27
Chicago, IL	7055	5965	5338	4747	3865	3184	2468	1204	645	288
New Orleans, LA	3464	2754	2102	1462	776	207	64	11	1	0
Boston, MA	6909	6062	5156	4412	3308	2454	1612	1046	486	217
Detroit, MI	6663	5904	5176	4558	3689	2988	2143	1361	755	345
St. Louis, MO	5499	4827	4220	3662	2775	2102	1381	806	418	194
New York, NY	6283	5517	4776	4048	2862	2022	1260	764	302	118
Akron, OH	6476	5708	4949	4287	3402	2734	1949	1266	779	309
Tulsa, OK	5087	4495	3892	3245	2205	1528	935	522	252	83
Portland, OR	7959	6754	5196	3809	2161	949	339	138	49	10
Philadelphia, PA	5488	5230	4536	3867	2846	2077	1296	722	323	128
Providence, RI	6584	5743	4982	4244	3238	2432	1723	1085	523	236
Columbia, SC	4326	3577	2913	2205	1411	870	451	203	55	10
Nashville, TN	4879	4162	3516	2868	2058	1401	836	436	197	74
Houston, TX	3104	2402	1808	1254	544	232	87	26	5	3
Salt Lake City, UT	7479	6812	6076	5173	3786	2877	1872	1095	556	192
Seattle, WA	7783	6747	5135	3623	2377	1057	356	143	54	20

*Compiled from summaries of hourly observations, NOAA, Ashville, N.C. (Independent summary for each city)

Experimental work in this area is expected over the next few years. Cooling ponds are more easily controlled than cooling towers with a much lower chance of major freezing problems.

Water in the towers may not use inhibitors, etc., if the water is injected into aquifers. A wet tower and cooling pond will provide a supply of oxygen and nutrients to the water. Davison (3), using a cooling pond, had success in removing the nutrients with a sand filter. No clogging problems occurred from particles or from biological growth in the aquifer. He anticipated problems with the water iron content and a possible growth of iron algae; however, the iron precipitated out in the cooling process and was removed in the filter. Schaetzle, (author of this book), has had some major problems with iron algae in a well system circulating water for heat-pump cooling.

An alternative to a cooling tower or pond is using a cold stream or river in parts of the country. Many streams attain near freezing temperatures or even freeze over in some cases. This water can be injected into an aquifer and an equal amount discarded. The injected water requires filtering and chlorination in some areas. If water from the aquifer is used for potable water additional measures are required. To satisfy environmental requirements, a heat exchanger can be used between river or lake water and aquifer water. An environmental study is required in each individual application.

In general, low-temperature cooling tower literature is extremely limited. However, a seminar on low-temperature cooling tower operation was presented at the 1980 Winter ASHRAE meeting in Los Angeles. Some temperature operation and safety procedures were discussed, but not in detail. The general content basically identifed problems which appeared during low-temperature operation. Information in the low-temperature cooling tower area is expected to expand rapidly in the next decade. Data are also required for direct winter "free cooling" systems which are now coming into existence in many applications.

System Performance

System performance is defined as the ratio of the attained cooling to energy required to attain the cooling or:

$$COP_{Cooling} = Cooling\ Effect/Input\ Energy \qquad (11-1)$$

For this application the cooling effect is the heat transfer to the cooling water. Air handler energy input is not substracted as for packaged airconditioning or heat-pump units. The heat-transfer rate is the water mass flow rate through the heat exchanger times the specific heat times the temperature change or:

$$Cooling\ Effect\ =\ q\ =\ m\ Cp\ (T_i - T_o) \qquad (11-2)$$

where "i" and "o" are subscripts respectively for in and out.

The quantity of water flow required for various cooling capacities over two temperature differences are:

	T = 10°F(5.5°C)		T = 20°F(11°C)	
1 Ton	2.4 gpm	0.545 m³/hr	1.2 gpm	0.272 m³/hr
10 Ton	24 gpm	5.45 m³/hr	12 gpm	2.72 m³/hr
100 Ton	240 gpm	54.5 m³/hr	120 gpm	27.2 m³/hr
150 Ton	360 gpm	81.7 m³/hr	180 gpm	40.0 m³/hr
200 Ton	480 gpm	109 m³/hr	240 gpm	54.4 m³/hr
300 Ton	720 gpm	163 m³/hr	360 gpm	81.7 m³/hr
500 Ton	1200 gpm	272 m³/hr	600 gpm	136 m³/hr

In general, commercial-size wells range in yield from 50 gpm (11.3 m^3/hr) to over 1,000 gpm (227 m^3/hr). Most of the potential applications proposed by the authors have used wells with capacities of 200 gpm. A minimum of two well pairs is normally recommended for reliability. Economics dictate one well pair in some applications.

The input energy includes all work required to chill the water, store the water, and transport the water through the chilling heat exchanger. The work does not include secondary loops; for example, the work required to operate a secondary loop of chilled water if a water-to-water heat exchanger is utilized. The work includes the following:

1. Work required to pump the water from the aquifer in order to cool the water. This includes the head loss between the wells in the aquifer, the head loss from the normal water table level to the cooling tower level, and the head loss in the pipes from the aquifer and return to the aquifer.

2. Work required to operate the cooling tower and to condition the water. This includes the head losses to the top of the cooling tower and the head loss across any spray nozzles in the cooling tower or pond. In order to obtain the required temperature level, the water might have to be cycled a number of times. The head loss in any filters utilized during the process must be included.

3. Head losses occur as the water is pumped from the aquifer through the cooling heat exchanger and returned to the aquifer. These include the head loss through the aquifer, the head loss from normal groundwater level to the heat exchanger (exchangers), and the head losses in the pipes from and to the aquifer. There will be short periods during low-temperature weather when water will circulate directly between the cooling tower and heat exchanger. This will decrease the input energy required but the decrease is expected to be negligible.

To give an insight into the performance of this system some example numbers are calculated, in sequence, for the above losses. It must be noted that each case is site and system specific.

A. Cooling the water and storing the water in the aquifer
Height from water table to cooling tower	50 ft
Head loss between wells	15 ft
Head loss in piping	30 ft
Total head loss in chilling water	95 ft

B. Losses in Filter and Cooling Tower
Head loss in cooling tower	15 ft
Head loss in filter	30 ft
Total head loss in cooling tower	45 ft

C. Cooling cycle head losses
Height from water table to cycle heat exchange	50 ft
Head loss between wells	15 ft
Head loss in piping	30 ft
Total head loss in cooling cycle	95 ft

Overall head loss	235 ft

To pump one pound of water through this load loss with a pump efficiency of 0.5 requires

$$W = 1 \text{ lb} \times 235 \text{ ft}/778 \text{ ft lb/Btu}/0.5 = 0.60 \text{ Btu}$$

Over a temperature difference of 20°F (7°C) the energy absorbed by the water is 20 Btu (21 KJ). The coefficient of performance is (Eqn. 11-1):

$$COP_{Cooling} = 20/0.6 = 33.3$$

or 33.3 Btu of cooling occurs for each Btu of energy input. In many systems this number can be more than doubled. The height from the water table to the ground level cooling tower for, as an example most of The Univeristy of Alabama campus, is only 25 ft. The head loss for the tower and filter is less than 20 feet in many cases. Piping losses, when the tower, cooling tower, and cooling system are located ajacent to each other, can be less than ten feet. In turn the losses could be higher. In all cases the coefficient of performance of 33, even with a 50 percent decrease, saves a great deal of energy over present central chiller systems with a coefficient of performance range of 3.0 to 4.0. Savings vary from 70 percent to more than 90 percent of present energy utilization.

System Capital Costs

System capital costs are in the same general range for direct cooling versus the standard chilling system. Each system is site dependent. Preliminary calculations for a number of buildings on The University of Alabama campus indicate the direct cooling capital costs are less than costs of the standard chilling system. However, the campus overlies an aquifer 70 ft thick with the water table 20 ft below the surface. For deeper aquifers the capital cost will be higher than that of standard chiller systems. The cooling tower cost in either system should be about equal, depending on climate and peak load. Direct cooling rejects less thermal energy, approximately 25 percent less, than the chiller system.

Costs were estimated (March 1980) for retrofitting a building with direct cooling systems. The present system is a 170 ton chiller. Estimated engineering costs by a professional engineer were:

Four wells with casing and sand screen (PVC) (This is a fixed quote by a well driller for six-inch wells)	$ 2,500
1,000 ft of six-inch pipe installed at $11.50/ft	11,500
Four pumping systems (fifteen HP) at list price (Estimate pumps could be acquired at 80% of list price installed, including pump controls)	11,000
One Heat exchanger with connectors etc. installed	8,000
System controls installed	3,000
	$ 36,000

This system gives 100 percent backup with the capability to supply 320 tons of cooling. Smaller pumping systems can be utilized at a cost savings of $2,600. Cost of repair of the present system, compressor failure, etc., was $45,000 during the year before the above calculations were made. Retrofit, depending on cooling use, will recover the investment in one year.

An additional indirect capital cost is power plant construction. For every ton of cooling with direct cooling, the peak power requirement is reduced approximately 750 watts. For the above installation, the reduction in peak power requirement (170 tons) is 125 Kw. At $800/kw capital cost, this amounts to $100,000.

Summary

The use of direct cooling with aquifer thermal energy storage is considered feasible based on energy savings and initial capital cost at 1980 prices. Energy savings of well over 70 percent for cooling are a reality. The system is geographically (climate) dependent and site (aquifer) dependent.

References

1. USEPA, "Interim Drinking Water Regulations--Control of Organic Chemical
    Contaminants in Drinking Water," 43 Fed. Reg. 5756, (Feb. 9, 1978).

2. <u>Cooling Tower Fundamentals and Applications Principles</u>, The Marley Company,
    Kansas City, (1969).

3. Davison, R.R., Harris, W.B., and Reddel, D.L., "Texas A & M Cold Water Aquifer
    Storage Project," in <u>Proceedings of Solar Energy Storage Options</u>,
    NTIS NO. CONF-790328-P2 Trinity University, (1979).

CHAPTER XII

OTHER APPLICATIONS

Numerous applications are available for aquifer thermal energy storage. The following sections consider some of the problems that relate to high-temperature storage. A brief discussion is included on district heating and cooling, and on greenhouse heating.

Process Heat Storage at Temperatures Above Boiling

When temperatures are above the atmospheric pressure boiling point, water must be pressurized if boiling is to be prevented. Boiling in an aquifer would cause immediate stratification of vapor to the top of the aquifer. Pressure at the bottom of the well, in the aquifer, is equal to the surface pressure plus pressure due to the head from the surface to aquifer depth. To maintain this pressure in a normal system requires continuous water injection. Higher pressure could rupture the aquifer. If the pressure is not maintained the water will boil. Some problems exist in high-temperature aquifer thermal energy storage. A system to allow high-temperature energy storage has been developed (1,2) and is discussed below.

An aquifer used for high-temperature storage must have a natural pressure greater than the saturation pressure at that temperature to prevent boiling. This requires that the natural head, the distance from the top of the aquifer to the piezometric surface, must be equivalent to the saturation pressure. Figure 12.1 illustrates the minimum height. The aquifer depth below the natural water table required to prevent boiling is shown in Fig. 12.2. Head is calculated to match the saturation pressure or:

$$h_{min} = P/\gamma \qquad\qquad (12\text{-}1)$$

Fluid pressure in the pipe extending into the aquifer, at aquifer level, will equal fluid pressure in pipe at the surface plus pressure due to the head (pressure) resulting from the column height from the surface to the well. This pressure must be reduced to aquifer pressure level. This is also a requirement for deep wells where the piezometric surface is well below the ground surface. Two methods are proposed:

> 1. Place a throttling valve (pressure regulator) in the pipe system to reduce pressure.

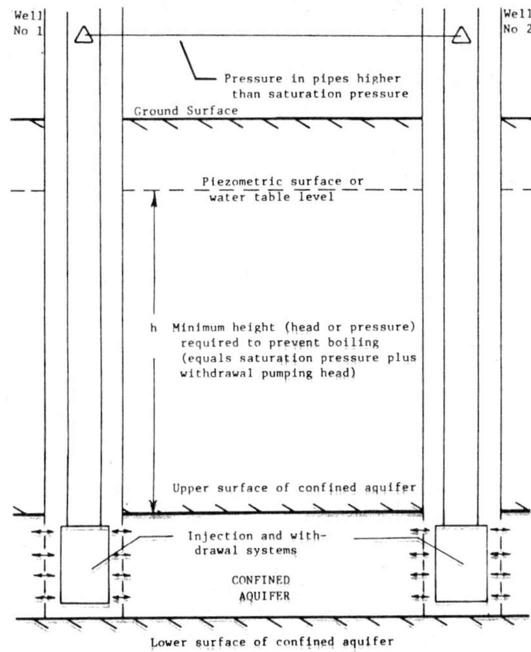

Fig. 12.1. Schematic of high temperature storage system.

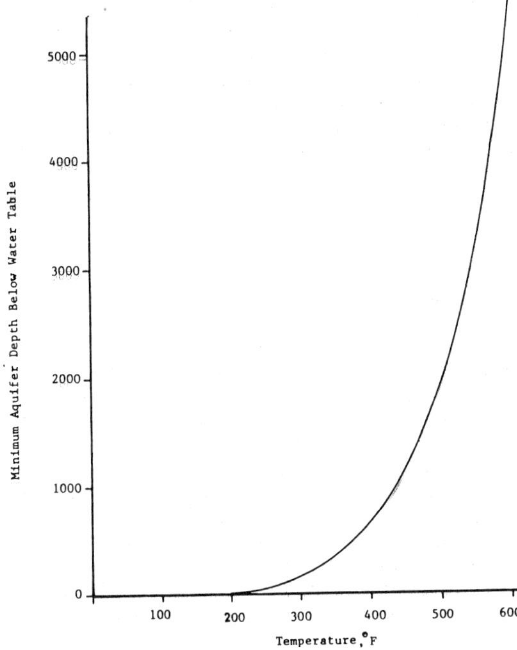

Fig. 12.2. Minimum aquifer depth below water table vs. temperature.

2. Place a turbine-generator system in the pipe at the aquifer level to recover work from the injected liquid. A turbine-generator combination is a potential solution. Both of these systems will reduce the well-pipe pressure to aquifer pressure. The second system is preferable based on energy requirements. The water being injected has already been pressurized from well level to the surface pressure. This pressure equals ΔP or:

$$\Delta P = P_{Saturation} + P_{Safety} + \qquad (12\text{-}2)$$

$$P_{Head \ (Well \ to \ Surface)}$$

Upon injection the work required to pump the water to this pressure is completely lost with a throttling valve system. Part of this work is recovered using a turbine-generator system, depending on efficiency (4). The work recovered is the available work times the turbine motor combination efficiency. This equals

$$W = \Delta P v \eta_t \qquad (12\text{-}3)$$

The quantity of work per pound-mass of water required to pump the water to the required pressure, P, is:

$$W = \frac{\Delta P v}{\eta_p} \qquad (12\text{-}4)$$

where η_p is the efficiency of the motor pump combination.

To remove the thermal energy, the flow is reversed and a pump (pumps) in the aquifer must pressurize the fluid to the pressure difference. The surface pressure normally must be greater than temperature saturation pressure in order to prevent boiling. This pressure is not necessary in some cases where flashing at the surface is desired. The work is normally the pump work noted in the previous equation.

Work required for the injection-recovery system is calculated with regard to the two different types of injection-recovery systems:

150

1. A throttling valve (pressure regulator) to reduce pressure for injection.

2. A motor-pump turbine-generator combination that recovers part of the work lost in injection and utilizes this work in recovery.

Work involved in the first system is twice the work required to pump water to the required pressure, once for storage and once for recovery.
Work required is given by:

$$W = \frac{2\Delta Pv}{\eta_p} \qquad (12\text{-}5)$$

With turbine-generator injection, part of the ideal work is recovered in each injection sequence. In this case:

$$W = 2\Delta Pv \left(\frac{1}{\eta_p} - \eta_t\right) \qquad (12\text{-}6)$$

Calculations have been made to give work required per unit of energy storage. For these calculations, temperature differences utilized are $100^\circ F$ ($38^\circ C$) and $200^\circ F$ ($93^\circ C$). The turbine-generator injection system and motor-pump efficiencies are assumed equal and curves are calculated for 40 percent, 60 percent and 80 percent efficiencies. These results are shown in Figs. 12.3 through 12.6. The first two figures are for throttling-value injection and the last two figures are for turbine-generator injection. Systems which perform as a motor-pump and turbine-generator are available. Energy savings for the 40 percent system (sixteen percent savings) probably would not justify the turbine-generator system. However, at 80 percent efficiency (64 percent savings) a motor-generator unit might be feasible. Efficiencies up to 90 percent are available at flow rates above 1000 gpm. Work increases rapidly with temperature increase.
The injection-recovery system allows storage of thermal energy at temperatures well above the normal atmospheric boiling temperatures. Storing thermal energy at temperatures over $212^\circ F$ ($100^\circ C$) is practical with this system. The system may also be applicable for deep wells (over 100 ft) at normal temperatures.
Numerous process applications are available for energy in this temperature range. In some applications process steam is required. Solar energy power generation discussed in Chapter X is an example. The steam cannot be generated at the storage temperature.

<u>Process Steam Generation</u>

Process steam can be generated a number of ways. The simplest is to throttle the hot withdrawal water to the desired pressure and corresponding temperature. The second method is to pass the withdrawal water through a steam generator. Two versions of steam generators will be discussed. The lower the steam temperature or pressure, the larger is the energy per unit-mass of water.
In the throttling process the withdrawal water is flashed into steam. This process is a constant enthalpy process. The wet steam can be used directly or the vapor and water can be separated and dry saturated steam will result. The following example gives an indication of the variations in quantity available. All water properties are from Keenan and Keyes Steam Tables.
Example: Water at $500^\circ F$ and 100 psi is withdrawn from an aquifer. Determine the quantity of saturated vapor which is produced at $450^\circ F$, $400^\circ F$, $350^\circ F$, $300^\circ F$ and $250^\circ F$.

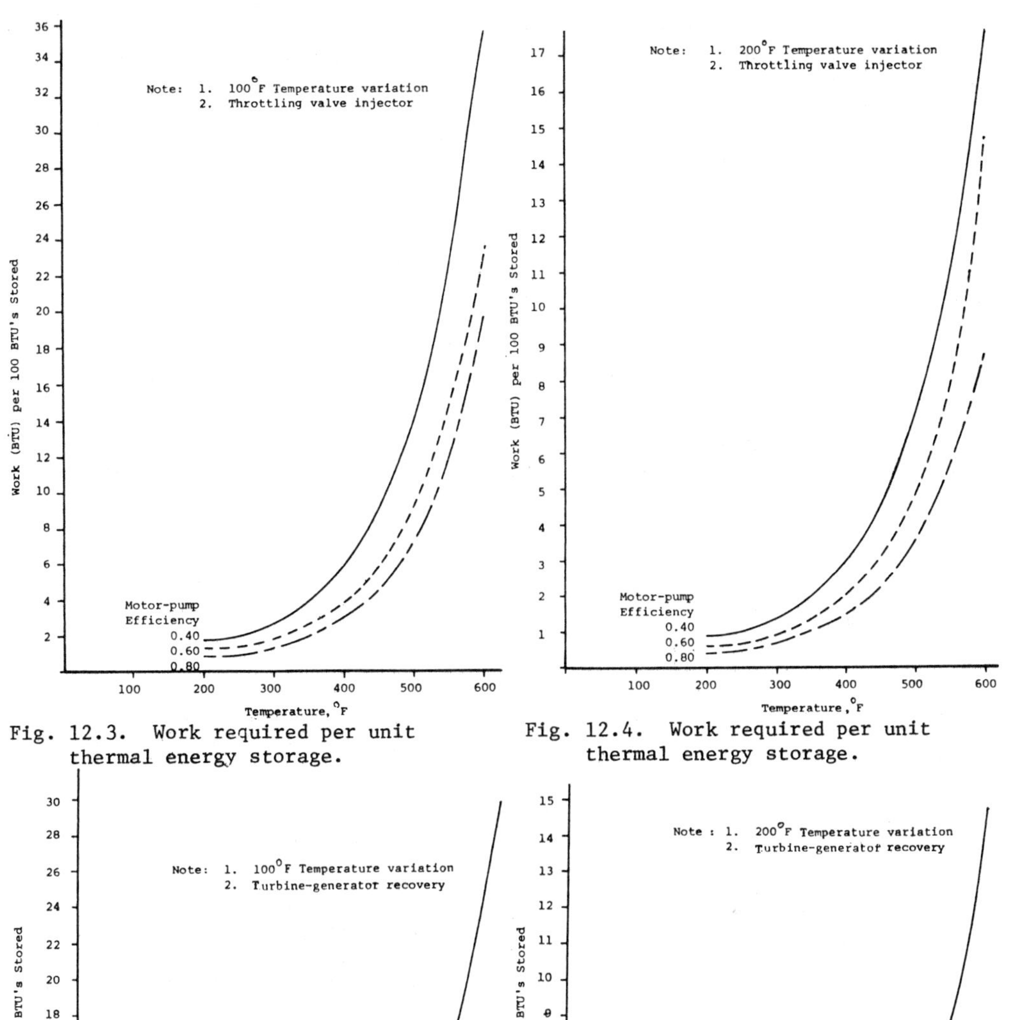

Fig. 12.3. Work required per unit thermal energy storage.

Fig. 12.4. Work required per unit thermal energy storage.

Fig. 12.5. Work required per unit thermal energy storage.

Fig. 12.6. Work required per unit thermal energy storage

152

Solution: The enthalpy at 500°F and 700 psia is 487.8 Btu/lbm.   The enthalpy at each of the above temperatures must match this value.   The enthalpy in the saturation region is:

$$h = (1 - x) h_f + x h_g$$

or:

$$x = \frac{h - h_f}{h_g - h_f}$$

(12-7)

where x is the fraction of mass that is vapor.   At 450°F:

$$x = \frac{487.8 - 430.1}{1204.6 - 430.1} = 0.0745$$

or 7.45 percent flashes to vapor (quality).
The values of x at the desired temperatures from 500°F water at 100 psi are:

°F	Quality
450	0.0745
400	0.136
350	0.191
300	0.240
250	0.285

Even at 250°F only 28.5 percent of the water flashes to vapor.

For steam generator heat transfer, the process is shown in Fig. 12.7 for the production of saturated vapor and in Fig. 12. 8 for the production of superheated steam.

Figure 12.7 shows the withdrawal water is cooled from T_H and T_L while the water to produce steam is heated from T_C to the saturation temperature and then converted to vapor. The temperatue of water being converted to steam does not normally start at T_L.   Note that T_L must be higher than T_C.   Liquid heating can be done in a counterflow heat exchanger. Energy rejected from the withdrawal water is injected into the water being converted to steam. Making an energy balance where heat transfer quantities are equal (no energy losses) and assuming a constant specific heat for the withdrawal water gives:

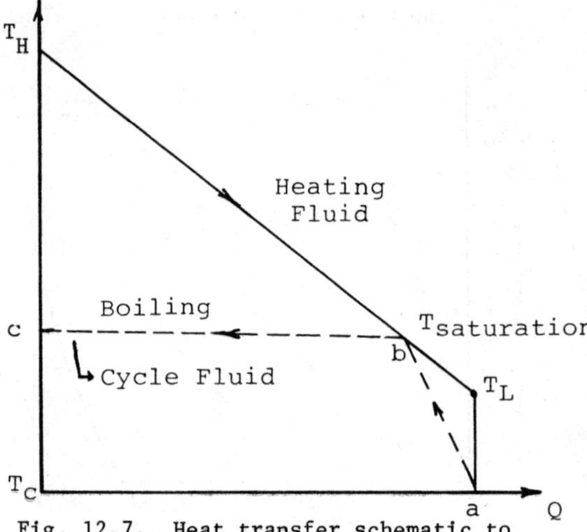

Fig. 12.7.   Heat transfer schematic to produce saturated steam.

$$m_{steam}(h_g - h_c) = m_{water} c_p (T_H - T_L)$$

(12-8)

or:

$$\frac{m_{water}}{m_{steam}} = \frac{h_g - h_c}{c_p (T_H - T_L)}$$

(12-9)

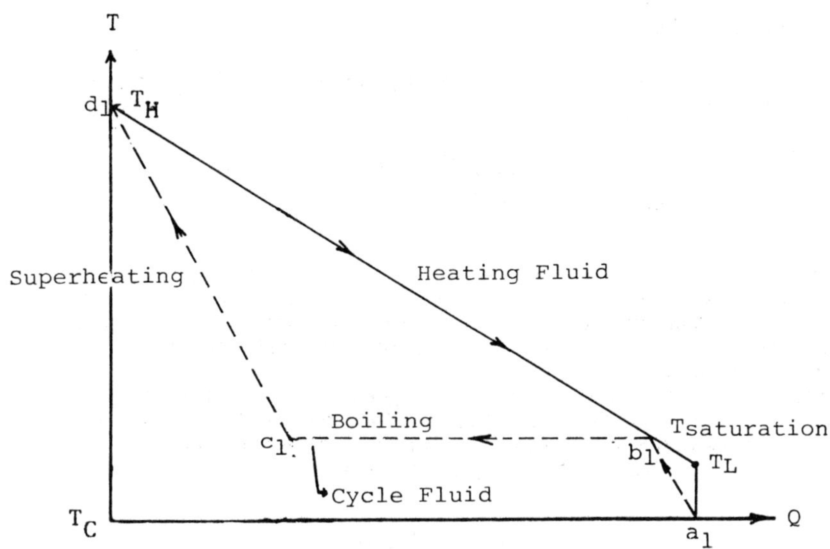

Fig. 12.8. Heat transfer schematic to produce superheated steam.

A second condition must also be met.   To satisfy the Second Law of
Thermodynamics energy transferred from the withdrawal water between T_H and
T_{Sat} must be at a higher temperature than or equal to boiling temperature. The
temperature of the remaining energy in the withdrawal water from T_{Sat} to T_L
must be greater than or equal to the temperature of the water to be vaporized in
the range of T_C to T_{Sat}.   Thermal energy transfers only from a high
temperature to a low temperature.   For each T_H and T_L there is only one
pressure where all the energy can be used. At this point an energy balance (Fig.
12.7) where all energy is utilized gives:

$$m_{water}c_p(T_H - T_{Sat}) = m_{steam}(h_g - h_f) \qquad (12\text{-}10)$$

and:

$$m_{water}c_p(T_{Sat} - T_L) = m_{steam}(h_f - h_c) \qquad (12\text{-}11)$$

Dividing the above equations results in:

$$\frac{T_H - T_{Sat}}{T_{Sat} - T_L} = \frac{h_g - h_f}{h_f - h_c} \qquad (12\text{-}12)$$

This ratio must be solved by trial and error to determine the saturation pressure
which uses all the available energy in the withdrawal water temperature range. As
T_L increases the saturation pressure increases and vice versa.
   For the production of superheated steam, the diagram in Fig. 12.8 is applicable.
Again applying an energy balance gives:

154

$$\frac{m_{water}}{m_{steam}} = \frac{h_{superheat} - h_c}{c_p(T_H - T_L)} \qquad (12\text{-}13)$$

Again a Second Law analysis (thermal energy only flows from a high temperature to a low temperature) gives:

$$\frac{m_{water}c_p(T_H - T_{Sat})}{m_{steam}c_p(T_{Sat} - T_L)} = \frac{m_{steam}(h - h_f)}{m_{steam}(h_f - h_c)} \qquad (12\text{-}14)$$

Again, only one saturation pressure and temperature satisfy the above equation. The solution is by trial and error. The final superheated steam temperature is T_H if an ideal counterflow heat exchanger is used for superheating. The saturation pressure for this system is lower than that for the saturation vapor system for the identical T_H-T_L.

Even though the derivation of the steam generation equations is for water, the enthalpies for any fluid can be substituted for the steam values. Freon, mercury and ammonia systems are examples. The First and Second Laws of Thermodynamics must be considered in design of hot water-vapor systems.

District Heating

District heating is an area where waste energy discarded over the entire year can be stored and used as required. In general storage in this application uses water below the boiling temperature. As a result heating must also be with water rather than with steam. Since water (liquid) is about 1600 times as dense as steam at 14.7 psi, the pipelines carry about the same amount of energy with water or steam. This is valid with steam up to 100 psi. The heat exchanger area must be larger as the convection heat transfer coefficient for water is much lower than the value for condensation.

Annual storage is located in the local area where the heating is required. A two-pipe insulated distribution system is used with the well pair storage system. Energy for storage can come from any waste energy source such as a power plant, laundry, restaurant, etc. Energy can be provided by boilers during off-peak energy periods.

Numerous advantages result from use of a large annual storage system. Some of these are:

1. Energy is stored locally. Local storage provides peak-load capacity which permits smaller pipelines from the energy source to energy user. A small thirty-mile pipeline transmitting energy from a power plant (at ten percent of peak load) could be used to continuously charge the aquifer. The only breakdown which affects heating is a local breakdown. Parallel community systems can normally take over in case of a breakdown.

2. Storage is on an annual storage basis. This allows energy injection on an annual basis allowing slow change and rapid discharge. Fuel for hot water storage such as natural gas which is normally in short supply in the winter·can be used during off-peak seasons at lower rates. A power plant or boiler providing the energy can be taken off line for periods without affecting the heating. Even solar energy collected during summer periods can be utilized.

3. The storage volume is a large system. This eliminates the need for backup equipment as the storage provided acts as a backup system.

Economically, several advantages result. These include smaller major (charging) pipelines and smaller local pipelines as loads are divided. Charging can use annually available waste heat sources cutting fuel requirements. Charging equipment can be smaller as only average annual loads are required, rather than peak loads. This ratio can be over a factor of ten. No backup equipment is required. Low fuel rates during off-peak times can be utilized.

Many communities can convert to such a system, including residential areas. A small annual charging line transferring thermal energy to a local residential neighborhood can be feasible. Implementation could be similar to the route natural gas implementation took many years ago.

Greenhouse Applications

The greenhouse is an excellent example for application of large amounts of low-temperature storage. The energy can be fed into storage from low quality waste energy sources or from solar energy. A suggested temperature range is 90°F (32°C) to 120°F (49°C). This allows the water to circulate through an air heater (natural or forced) and then be used for plant watering. Temperature for watering purposes can be lowered to 80°F. Plants thrive better with warm water.

Figure 12.9 is a sketch illustrating operation of a simple solar system for a greenhouse which can be installed locally. For heating water a section of level ground is enclosed by 4 x 4 lumber. This could also be 2 x 4 lumber with stakes. A single sheet of black polyethelene is placed next to the ground and a clear sheet of polyethelene is spread on top of the sheet. The cold water is inserted on one end and warm water heated by solar energy is removed from the opposite end. Cold water is withdrawn from the cold well and warm water is injected into the hot well.

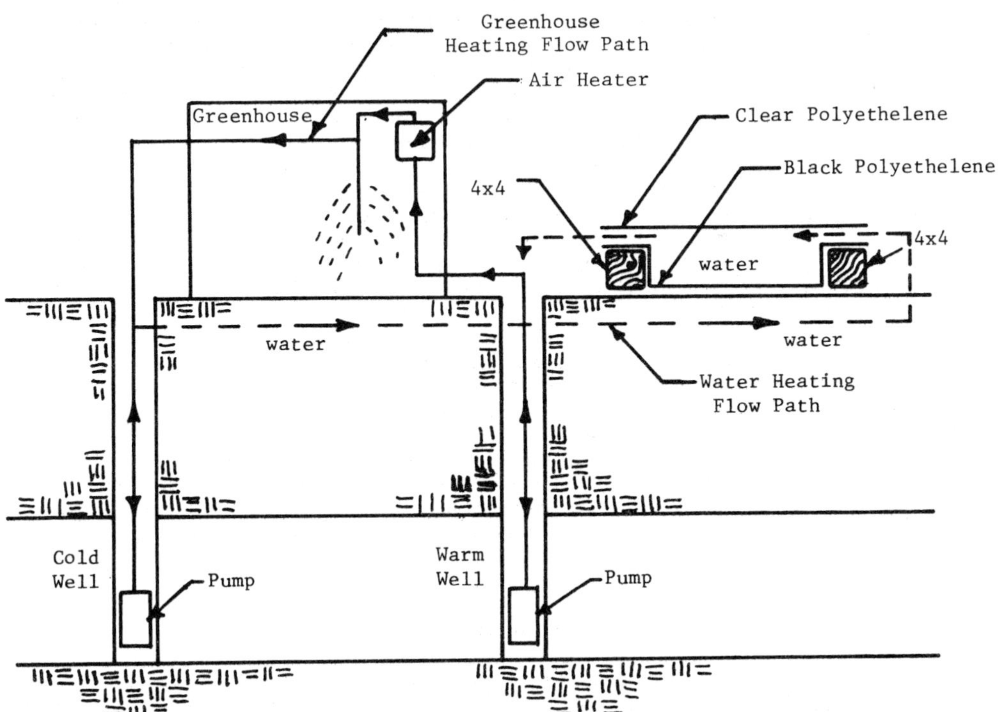

Fig. 12.9. Greenhouse application.

During the months of June through August the system is expected to collect approximately 1000 Btu/ft^2 per day or about 100,000 Btu/ft^2 per season. For an area of 100 ft x 100 ft this amounts to 1000 million Btus per year. Collection is site specific. The energy per year could be over twice the above quantity or slightly below the above quantity. If labor is not included, the cost of energy can be less than \$.50/million Btu with a one-year depreciation. Other types of manufactured swimming pool collectors can be utilized. Care should be taken to eliminate any contact with air for water to be injected into the aquifer.

When the warm water is withdrawn it can be put through an air heater. The heater can be natural convection or forced convection. For additional heating the semi-warm water can be used for watering. This not only helps the heating but also increases plant growth. The system provides an economical method of heating for the hothouse farmer. Most of the labor can be performed directly by the farmer.

Heat Exchanger Fouling

In many present day water systems, water is circulated in a closed loop. The water is cooled in a cooling tower and heated by waste heat or in some cases directly from other energy sources. Mineral content and pH are closely controlled. In the proposed system, water in the aquifers and mineral additions during water circulation must be evaluated. Chemicals cannot be added.

Knudson has been active in condenser water fouling research for many years (3,4,5,6). His basic premise is that mineral buildup along the heat exchanger surfaces occurs continuously. After a period of time the mineral buildup rate and erosion rate reach equilibrium. This is demonstrated by plotting fouling resistance "R_f", the change in the heat transfer coefficient, versus time. In general the time to equilibrium varies from 150 to 600 hours. Typical curves are shown in Fig. 12.10 for hardness levels around 50 ppm at different locations.

Knudson's series of papers have shown the variations for a number of minerals. Knudson's test data and that for many aquifers cannot be compared directly; however, maximum hardness values are of similar magnitude.

In his papers, Knudson's maximum value of resistance fouling is of the order of magnitude of 0.0004 ft^2 hr ^{o}F/Btu. The value for sea water is given as 0.0005 ft^2 hr ^{o}F/Btu (7). The effect of this level is shown to be minor (5,6).

The general range of heat transfer coefficients for a Freon 12 condenser is 80 to 150 Btu/ft^2 hr ^{o}F. The reciprocal or resistance therefore varies from 0.012 to 0.00667 ft^2 hr ^{o}F/Btu. The change due to mineral fouling is of the order of a few percent.

Possibilities of heat-exchanger fouling and aquifer clogging resulting from biological growth must be evaluated in the laboratory or in demonstration projects. These variables can be controlled to an extent, if necessary, by the addition of chemical inhibitors. Another possibility is to eliminate all air-water interfaces in the system.

Fouling will lower heat transfer coefficients by a few percent. This must be taken into account in the design of systems by using larger heat exchangers. The increase will be a function of mineral content in the aquifer water. Overall problems with the fouling of heat exchangers on the aquifer side is expected to be minimal.

Summary

As may be seen from the examples noted in the last few chapters, numerous economical applications exist for aquifer thermal energy systems. The large annual thermal energy storage system provided by nature is much more economical than anything man can produce.

References

1. Schaetzle, W.J., Brett, C.E., and Ansari, J.M., "Solar Energy Storage in Aquifers for a Solar Power Plant," _Proceedings of the 2nd Miami International Conference on Alternative Energy Sources_, University of Miami, (Dec. 1979).

2. Ansari, J.M., _High Temperature Thermal Energy Storage in Aquifers with a Solar Power Plant Application_, Ph.D. Dissertation, The University of Alabama (1980).

3. Knudson, J.G., Personal Communication to Dr. Schaetzle (Ten fouling curves), (June, 1978).

4. Morse, R.W., and Knudson, J.G., "Effect of Alkalinity on the Scaling of Simulated Cooling Tower Water," _The Canadian Journal of Chemical Engineering_, Vol. 55 (June, 1977).

5. Knudson, J.G., and Story, M., "The Effect of Heat Transfer Surface Temperature on the Scaling Behavior of Simulated Cooling Tower Water," _AICHE Symposium Series_, No. 174, Vol. 74 (1977).

6. Lee, S.H. and Knudson, J.G., _Scaling Characteristics of Cooling Tower Water_, Department of Chemical Engineering, Oregon State, (August, 1978).

7. Holman, J.P., _Heat Transfer_, McGraw Hill (1976).

CHAPTER XIII

ENGINEERING ECONOMICS

INTRODUCTION

The magnitude of the investment required to design and install a system for thermal energy storage in aquifers demands a thorough economic evaluation and justification. The large initial investment and extended benefit period associated with such systems make the analysis methods commonly known as "engineering economics" appropriate. Engineering economics provides the means to compare the relative economic merits of alternative systems in a variety of situations.

The principles of engineering economics are widely practiced by corporate engineers making recommendations on large capital investment projects. For such engineers this chapter presents only new applications of familiar techniques. Systems involving thermal energy storage in aquifers have considerable potential application in the small business, governmental, and private consumer sectors. Much decision making within those sectors is made without the benefit of engineering economics techniques. This chapter is intended to introduce planners and analysts from those sectors to the more generally applied engineering economic principles. A pair of case studies demonstrate engineering economics application to systems involving thermal energy storage in aquifers.

The structure of this chapter follows the general phases used in the economic justification of most major systems:

    1. Estimation of system cost and benefits
    2. Engineering economic analysis
    3. Sensitivity analysis
    4. Selection of the "best" alternative

A pair of case studies are presented to illustrate the methodologies employed. The chapter concludes with a summary of the uncertainties inherent in such an analysis. A list of references is provided with specific entries suggested for the reader not familiar with engineering economics techniques.

System Cost and Benefits Estimation

System costs are typically of two types--the initial capital investment which must be amortized or depreciated over the life of the system and the annual cost to operate and maintain the system. The successful engineering economics analysis is highly dependent on accurate estimates of the proposed system's cost and benefits. Cost engineering is an area involved with estimation and control of the system cost. Such estimates are normally based on past experience which, for the most part, is lacking for systems involving thermal energy storage in aquifers. However, many contractors and engineers are familiar with the major system components.

System	Major Components
Water Supply	Wells, pumps and storage tanks
Water Distribution	Piping and pumps
Energy Utilization	Water source heat pumps, solar collectors, cooling towers and chillers

Initial purchase cost and operating energy consumption data are available for each of these major components. System integrating and control costs are more difficult to estimate and require experienced engineering judgement.

Energy conservation is the major system benefit and generally is estimated relative to the current system's energy usage or that of a proposed alternative system. Reasonable estimates of the energy-savings potential can be made based on thermal loads and efficiency ratings. At present, the most difficult aspect of benefits estimation is the forecasting of future energy costs. Lacking any unforeseen economic crisis or new technological breakthroughs, energy costs appear to be headed for increases substantially above the increase in total consumer prices. This point is a strong economic incentive for the consideration of systems involving thermal energy storage in aquifers.

Cost Engineering

The general term "Cost Engineering" has been developed to include the task of cost and benefits estimation. Cost engineering is defined by the American Association of Cost Engineers (AACE) as "that area of engineering practice where engineering judgement and experience are utilized in the application of scientific principles and techniques to the problems of cost estimation, cost control and profitability." While this definition clearly fits the problem at hand little formal literature exists to guide its application to a new area, such as, thermal energy storage in aquifers. A recent book by Clark and Lorenzoni (4) entitled Applied Cost Engineering provides a background for the cost engineer associated with large scale oil refinery projects. While thermal energy storage in aquifer applications are likely to be on a smaller scale, the general methods covered are applicable. For example both refineries and aquifer storage systems include piping networks. An older book by Page (12) specifically covers the cost of pumping systems. Inflation has made the numerous charts useless except as means to compare the relative cost of similar equipment over a range of capacities. The six-tenth factor rule is suggested to estimate the cost of equipment with a capacity not included in available price lists:

$$C_y = 0.6 \ C_x \ (Y/X) \eqno(13-1$$

where C_y is the estimated cost of a unit with capacity Y and C_x is the known cost of a unit with capacity X. The six-tenth factor rule is valuable in the rough cost estimation for systems involving new capacity components.

Capital Costs

The engineering economist uses the term capital cost to refer to those system costs which for tax purposes must be written off or recovered by depreciation. Long term costs such as purchase of land, building, durable equipment and patents are said to be "capitalized" or carried on the books over their useful or economic life. A general quideline as to whether a cost is capitalized or not is the expected useful life of the asset. Items with an expected useful life of two or

160

more years generally must be capitalized. Some expenditures such as engineering analysis, computer programs, and wells which do not produce sufficient water are difficult to characterize as capital or non-capital costs. Such cases should be left to the tax advisors.

The cost of these capital investments must be carefully estimated to assure that the initial commitment of funds is sufficient to carry the project through its design and construction and into its payback period. Systems using thermal energy storage in aquifers are likely to involve three subsystems with a significant capital investment.

>   Water Supply, wells and pumps (see Chapter V & VI)
>   Water Distribution, piping network (see Chapter VII)
>   Energy Utilization, heat pumps, etc., (see Chapter VIII, IX, and X)

Commercial business and private investors should be aware of the income tax incentives available from the Federal government. The business tax credit provides for an additional ten percent investment tax credit (ITC) over and above the ten percent conventional ITC for approved solar expenditures in commercial applications. Only the ten percent ITC is currently available for commercial systems involving thermal energy storage in aquifers without solar equipment. Private residential systems involving solar equipment are entitled to a residential tax credit of 30 percent of the first $2,000 and 20 percent of the next $8,000 of expenditures for solar equipment, up to a total maximum credit of $2,200. Future changes in the Federal tax code may include aquifer storage systems as an incentive for energy conservation. Many states also offer investment and energy conservation tax credits. Such credits are too complex and frequently changed to be tabulated here. The engineering economist should be aware of the appropriate state and local tax codes.

Annual Cost and Benefits

All systems have day-to-day operating and maintenance expenses which must be paid to enjoy the benefits expected when the original capital investment was made. These costs are typically collected or tabulated on an annual basis because many costs, such as taxes or insurance, are payable once per year. The annual cycle also matches the budget or income and expense reporting cycle of most organizations.

The annual costs associated with systems involving thermal energy storage in aquifers include:

>   Energy
>   Insurance
>   Interest on loans
>   Maintenance
>   Repairs or replacement of broken equipment
>   Periodic or preventive maintenance
>   Supervision or Management
>   Taxes
>   Income
>   Property

These annual costs are expected to be offset by benefits when compared to the present system or some alternative system which may be installed in place of the proposed system. Energy savings are likely to be the major benefit obtained from a system using thermal energy storage in aquifers. Other benefits might also include:

Reduced maintenance due to newer and/or simpler equipment.
Reduced income tax, due to increased deductions for property tax, depreciation and interest payments.
Increased public image for the organization because of the successful application of a novel system.
Increased experience gained through the new application of technology.

Generally, the annual costs and the first two benefits can be quantified based on engineering analysis and past experience. The last two benefits are intangibles which deserve consideration but usually must be left unquantified.

The system's energy usage or savings versus those of some base system can be calculated using system component efficiencies for a typical or normal heating and cooling season (annual cycle). The annual energy consumption or savings might also be computed for some extreme cases, such as exceptionally mild or severe seasons. The current uncertainties associated with future energy prices makes the conversion from energy units to dollars a risky task. Two methods of dealing with this problem are demonstrated in the case studies.

Annual system maintenance costs can be obtained from past experience with similar types of components. Bezdek, et al, (2, 9) used an annual maintenance cost of 1.5 percent of the capital investment for residential solar heating systems. Christian (3) suggests a five percent factor for water-source heat pumps. The actual cost for such maintenance is likely to depend on the equipment's age and the procedures developed for such maintenance. Residential heating and cooling systems often receive no maintenance until they fail to function. Large-scale commercial heating plants receive constant supervision and preventive maintenance. Economical preventive maintenance programs are likely to be developed in applications where scale warrants them.

The other annual cost factors can be estimated based on the organization's current tax and borrowing status. It should be noted that a single system might involve several different types of organizations and hence be subject to different tax rules and borrowing incentives or restrictions. For example, the supply and distribution of water may be controlled by a non-profit, tax-exempt corporation, the energy users may be public profit-making corporations or businesses, private consumers or governmental units.

Engineering Economic Analysis

The second economic justification phase uses the techniques of engineering economics to determine if the system's future benefits are sufficient to justify the required investment. Payback period is a simplistic approach used to evaluate short-term investment opportunities. Payback period is inadequate when alternative systems have differing expected useful lifetimes or when the time value of money must be considered. Two general approaches which consider the time value of money or interest rate are reviewed. First, the initial capital investment can be annualized using the capital recovery method to develop a uniform annual cost for the proposed system. The capital recovery method can not readily handle the special problems caused by income tax considerations and variable component lifetimes. The second method estimates the annual cash flow for the system and determines the rate of return on investment (ROI) generated on the annual cash flow. ROI analysis is difficult without the aid of a computer or sophisticated calculator but provides the most comprehensive economic justification tool and can be applied to any cash flow pattern.

The reader unfamiliar with the concept is directed to one or more of the many outstanding books on the subject of engineering economics (1, 5, 8, 11, 13, 16, 17, and 18). The Engineering Economist is a technical journal published quarterly by the Engineering Economy Divisions of the American Society for Engineering Education and the American Institute of Industrial Engineers. That journal is devoted to the problems of capital investment and frequently contains articles (6, 7) pertinent to the analysis of energy systems.

Each of the three engineering economic analysis techniques will be used to evaluate the relative economic merits of two example investment alternatives:

Alternative	Initial Investment	Annual Benefit	Useful Life (Years)
A	$10,000	$2,500	5
B	$10,000	$2,000	10

Payback Period

The most simplistic engineering economic analysis approach is to divide the system's initial capital cost by its expected annual net benefit (gross benefit less annual cost). The result is the payback period or the number of years required to recover the initial capital investment. This procedure is used to compare the economic merits of alternative systems or to screen potential capital projects. For example, an organization may terminate work following preliminary engineering design on all projects which show a payback greater than five years.

Now consider the two example investment alternatives. Alternative A has a payback period of $10,000/$2,500 = 4 years, while alternative B has a payback period of $10,000/$2,000 = 5 years. Thus, alternative A seems to be the most attractive investment. Clearly the payback method does not consider the fact that investment alternative B returns a benefit for twice as long (10 years) as alternative A (5 years).

The payback period method does not adequately handle investments with differing useful lifetimes and does not consider the time value of money. In spite of these limitations the payback period method is widely used in corporate decision making and is usually the first economic statistic used to analyze an investment problem.

Capital Recovery Method

The interest paid on borrowed funds or the potential interest which might be earned by not investing available funds in the proposed system represents the time value of money. High interest rates and extended project lifetimes make the time value of money an extremely significant factor affecting the decisions associated with major energy systems. The rule of 72 illustrates this point. The number of years, N, required for an investment to double in value or inflation to reduce purchasing power by 50 percent is approximately given by:

$$N = 72/I \quad \text{or} \quad NI = 72 \tag{13-2}$$

where I is the compounded interest or inflation rate. Thus, at twelve percent compound interest an investment will double in value in about six years. Likewise, ten percent inflation halves the value of a dollar in about 7.2 years.

While the rule of 72 presents a quick investment evaluation tool it can not match the precision of the capital recovery method. The capital recovery method is used to compute the annual equivalent capital cost. The capital recovery factor, CRF, assures the recovery of the initial investment with interest over the economic life of the system:

$$CRF(i\%,n) = i(1+i)^n/((1+i)^n -1) \tag{13-3}$$

where i is the annual interest rate and n is the system's economic life in years:

The capital recovery method can be used to annualize the cost of any investment. The annual investment cost can be added to the other annual costs. Only systems whose annual benefits exceed their total annual cost, operating and maintenance plus capital recovery, should be recommended from an economic point of view.

Application of the capital recovery method implies that the appropriate interest rate is known and fixed for the term of the investment.

Consider again our two example investment alternatives and assume ten percent to be an appropriate interest rate. The capital recovery factor for each alternative using 13-3 is:

Alternative A: CRF (10%, 5) = 0.10 $(1 + 0.10)^5/((1 + 0.10)^5-1)$ = 0.2638
Alternative B: CRF (10%, 10) = 0.10 $(1 + 0.10)^{10}/((1 + 0.10)^{10}-1)$ = 0.1627

The net annual benefit for each alternative is:

Alternative A: Net annual benefit = \$2,500 - \$10,000 x 0.2638 = - \$138
Alternative B: Net annual benefit = \$2,000 - \$10,000 x 0.1627 =   \$373

It is clear that alternative B is the superior investment having a positive annual benefit compared to a negative benefit for alternative A at ten percent interest. If the interest rate were to change, a new calculation would be required to see if alternative B is still profitable. For example, suppose the interest rate rose to twelve percent, CRF (12%,10) = 0.1770 and the net annual benefit = \$2,000 - \$10,000 x 0.1770 = \$230. Thus at twelve percent interest alternative B is still profitable. Investment A would have a net annual benefit of - \$274.

A complex system may involve several components with differing lifetimes. The annualized capital recovery factor must be computed for each component and their sum included in the total annual cost versus benefits comparison. Most engineering economics books include capital recovery factor tables for the more commonly used interest rate values. Equation 13-3 can also be used to compute directly the appropriate factor given the interest rate and lifetime. Notice that the interest rate is assumed to be fixed over the life of the system. The first case study illustrates in detail the application of the capital recovery method to a system involving thermal energy storage in aquifers.

Return on Investment
As interest rates change the user of the capital recovery method must recompute the CRF and the associated net annual benefit. The return on investment (ROI) method eliminates this problem by not assuming an interest rate but rather finds the interest rate at which the net annual benefit becomes zero. For investment alternative B find the value of i which makes:

$$\text{Net Annual Benefit} = \$2,000 - \$10,000 \text{ CRF } (i\%,10) = 0 \qquad (13\text{-}4)$$

By rewriting:

$$\$10,000 \text{ CRF } (i\%,10) = \$2,000$$
$$\text{CRF } (i\%,10) = 0.2000 \qquad (13\text{-}5)$$

Using a number of CRF tables and some luck one can find the appropriate value of the interest rate i or ROI. In this case CRF (15%,10) = 0.1993 which may be close enough for most engineering economic decision making. A more exact method of finding the ROI would substitute equation 13-3 for the CRF term into equation 13-5 yielding:

$$\text{CRF } (i\%,10) = i(1+i)^{10}/((1+i)^{10}-1) = 0.2000 \qquad (13\text{-}6)$$

Rearrange terms to yield:

$$(i-0.2)(1+i)^{10} = -0.2 \qquad (13\text{-}7)$$

Equation 13-7 could be solved for i but not without considerable difficulty.

164

An iterative search approach is usually used to find the ROI value for i by seeking values that make the net annual benefit closer to zero. A typical search for investment alternative B might follow the steps below:

i%	CRF (i%,10)	Net Annual Benefit	Result
10	0.1627	$373	i too small
18	0.2225	-$225	i too large
15	0.1993	$7	i slightly small
15.2	0.2008	-$8	i slightly large
15.1	0.2000	$0	correct ROI value

The ROI for investment alternative A is 7.9 percent. A FORTRAN computer subroutine which calculates the ROI rate for an arbituary cash flow using a binary, iterative search is included in Ref. 15.

The procedure for using the ROI method involves determining the system's annual cash flow. Cash flow is defined as the system's total benefits (gains) for a given year minus its total costs. The annual cash flow can be either positive, negative, or zero. Most investment problems begin with a large negative cash flow in year zero (the initial capital investment). For ROI analysis all cash flows are assumed to take place at the end of each analysis year. Thus, the initial system investment is made on the last day of year zero. The first benefit would be expected one year hence. This annual model is admittedly rather gross but has generally been found to be sufficiently accurate for a wide range of capital investment problems.

The cash flow analysis should include the factors previously considered in the section on Annual Cost and Benefits. A special note should be made on income tax paid by private corporations and individuals. If the system's owner pays business related income taxes, the system's capital cost should be depreciated over its useful year. This depreciation is recognized as business expense and hence is excluded from the firm's profits for income tax calculation. A rapid depreciation schedule is usually preferred to gain the cash flow advantage of reduced income taxes early in the system's life. Any of the engineering economics books cover the various tax and depreciation methods in detail.

Once the annual cash flow pattern has been developed the ROI can be found by a search technique. Because each cash flow is likely to be different in each year of the system's life, the capital recovery method discussed earlier is not appropriate, rather discounted cash flow must be used. The discounted cash flow method converts annual cash flows into an equivalent year zero cash flow amount. The appropriate present worth factor:

$$PWF(i\%,n) \;=\; (1+i)^{-n} \qquad\qquad (13\text{-}8)$$

is multiplied by the cash flow for year n. All of the discounted cash flows are then summed. The exact ROI has been found when the sum of the discounted cash flow is exactly zero. A positive sum indicates that the value of i is less than the true ROI. A negative sum indicates the opposite, i greater than ROI. Table 13.1 illustrates an ROI calculation for simple investment problem with non-uniform annual cash flows. The cash flow in this example started with a $10,000 investment (negative) in year zero. The net annual benefit grew from $2,000 to $5,000 over the four year useful life of the investment. The first non-negative cummulative cash flow occurs in year four indicating a four year payback period. The search of the ROI was started with i = 0.10. This yielded a cummulative discounted cash flow of $718. A positive final cummulative discounted cash flow indicates that the trial value of i was too low. The second iteration used 0.15 as the trial interest rate. This resulted in a negative final cummulative discounted cash flow. Hence the true ROI lies in the interval of 0.10 to 0.15. The third iteration used 0.13 which proved to be very close to the actual value of 0.129.

TABLE 13.1. ROI Calculations, Non-uniform Cash Flow

Year n	Annual Cash Flow	Cumm. Cash Flow	Discounted Cash Flow @ i = 0.10		
			PWF	Annual	Cumm.
0	-10,000	-10,000	1.000	-10,000	-10,000
1	2,000	- 8,000	0.909	1,818	- 8,182
2	3,000	- 5,000	0.826	2,479	- 5,702
3	4,000	- 1,000	0.751	3,005	- 2,697
4	5,000	4,000	0,683	3,415	718

Year n	Discounted Cash Flow @ i = 0.15			Discounted Cash Flow @ i = 0.13		
	PWF	Annual	Cumm.	PWF	Annual	Cumm.
0	1.000	-10,000	-10,000	1.000	-10,000	-10,000
1	0.870	1,739	- 8,261	0.885	1,770	- 8,230
2	0.756	2,268	- 5,993	0.783	2,349	- 5,880
3	0.658	2,630	- 3,363	0.693	2,772	- 3,108
4	0.572	2,859	- 504	0.613	3,067	- 42

Sensitivity Analysis

The results of an engineering economic analysis are subject to errors in the original estimates of system costs and benefits. Sensitivity analysis is a general technique used to determine the range over which those estimates can vary without changing the recommendation made from the analysis. Digital computer simulation is one means to test the sensitivity of a recommendation in an uncertain environment. An Energy System Investment Simulator (ESIS) program has been developed by Seppanen (15) to compute both the expected payback period and the ROI when estimation uncertainty is present.

Program ESIS uses "Monte-Carlo" simulation to randomly select the values for analysis parameters that are not known with certainty. Multiple runs of the ESIS program provides the engineering economist with the probable ranges for both the payback period and the ROI. The types of analysis parameters or data required by the ESIS program include:

External investment rate
Useful life of the investment (may be probabilistic)
Capital investment amount (may be probabilistic)
Capital investment timing (all at end of year zero or distributed
    over a ten year period)
Borrowed funds (repayment schedule, interest rate and fraction of initial
    investment borrowed)
System salvage value
Depreciation method and additional first year depreciation rate
Annual benefits (may be probabilistic and/or include an annual trend)
Annual expense cost (may be probabilistic)
Income tax and investment tax credit rates

Program ESIS functions by determining the system's annual cash flow and then the ROI. If one or more of the probabilistic parameters are specified, multiple runs are made and the output results statistically and graphically summarized.

166

Several ESIS features make it particularily valuable in the analysis of systems involving thermal energy storage in aquifers.
1. ESIS can handle the uncertainties in capital investment cost and annual benefits.
2. ESIS can accommodate a pattern of increasing annual benefits caused by energy cost increases above the general inflation rate.
3. ESIS can handle multiple investment tax credits, such as, those currently allowed for solar energy systems.

Alternative Selection

The final economic justification phase is the selection of the "best" alternative. The word best is used with some caution because of the previously mentioned uncertainties. Usually, economics alone are not sufficient to make the final selection decision. For example, the engineering economics study may narrow a field of ten potential systems to two or three closely ranked "superior" alternatives. In other cases only two alternatives may have been considered and the results are in the form of future scenarios which favor each alternative, i.e., choose system A if energy price increases are expected to be at least 15% per year, otherwise choose system B.

Thus, the final system decision must often be made with more than economic facts. Corporate or community image might well be a key decision factor in the acceptance or rejection of a novel energy system. Other factors might include; risk, potential for future markets, energy availability and availability of investment funds.

Case Studies

Two cases are presented to demonstrate the application of engineering economics techniques to the analysis of systems involving thermal energy storage in aquifers. Case 1 involves new residential construction where a water-source heat pump system using aquifer storage is compared to a more conventional air-source heat pump system for heating and cooling. The capital recovery method is used and a graphical sensitivity analysis is performed on the cost of electrical energy. Case 2 involves the conversion of a central city business district to an aquifer storage system. The ROI method is demonstrated. Both cases were originally presented by Schaetzle, _et al_, (14).

Case I, New Residential Construction

New residential construction on a subdivision or community basis represents an attractive setting for systems using thermal energy storage in aquifers. The layout of the community can be designed with consideration given to potential well sites, the water distribution system configuration and solar heat collection locations. A water-source heat pump based system may eliminate the need for a natural gas distribution system and may serve as the fire protection water source, thus reducing the required size of the potable water mains. While these factors seem to favor the proposed system in new residential construction, they are difficult to quantify and are likely to vary in impact from community to community. In addition, the proposed system is likely to favor all-electric communities which in turn will impact the demand pattern for electricity.

A comparative approach has been taken in the engineering economic analysis of alternative systems for new residential heating and cooling. A total annual cost has been computed for a single home using each of two systems (water-source heat pump using thermal energy storage in an aquifer and individual air-source heat pump) in each of three cities (Miami, Birmingham and Detroit) selected to demonstrate a range of heating and cooling loads. The total annual cost provides for capital recovery, system maintenance and energy costs. Because of the uncertainty associated with future energy costs, the analysis is based on a range of possible energy costs.

Table 13.2 illustrates the development of capital and maintenance costs for the alternative systems. The capital costs for the installed heating and cooling units are from Christian (3).

TABLE 13.2. Annualized Capital and Maintenance Cost

System Alternatives	Home Unit	Water Supply & Distribution Systems	Total	Annual Equivalent Cost	Annual Maintenance Cost
Air-Source Heat Pump	$ 2,800	$ 0	$2,800	$ 329	$ 42
Water-Source Heat Pump	2,000	1,391	3,391	398	51

1. Assume a 20-home module size, total capital cost $27,820 see table 13.3
2. Based on a 20-year economic life for the entire system, no salvage value, no tax consideration, 10% interest rate using capital recovery factor = 0.11746.
3. Annual Maintenance Cost = 1.5 % Total Capital Cost.

The capital costs for the installed water supply and distribution systems used by the proposed water-source heat pump system use are developed in Table 13.3 based on Means (10). The water supply system consists of a pair of wells, pumps and capacitance tanks with the necessary controls and solar collectors for

TABLE 13.3. Water Supply and Distribution System Capital Cost

Water Supply System

Well Construction (2 wells, less than 200 ft)	$3,000	
Pumping System (pump, motor and controls)	$6,000	
Capacitance Tank (2 tanks)	$ 260	
Solar Collectors (1000 ft at $3/ft)	$3,000	$12,260

Water Distribution System Using PVC Pipe

Double 6-inch, Primary (1000 ft at $9.20/ft)	$9,200	
Double 1.5-inch, Secondary (1000 ft at $3.40/ft)	$3,400	
Trench (2000 ft at $1.48/ft)	$2,960	$15,560

Total Water Supply and Distribution System Capital Cost	$27,820

Based on 100 ft average home spacing on both sides of the street with a 50 ft setback from street center line and a 20-home module.

additional heat makeup. The pumps are sized to service a 20-home module using an aquifer with 50 or more ft of porous and highly permeable water-bearing section within 200 ft of the surface. A less favorable aquifer could significantly increase the cost of the water supply system.

The water distribution system costs are based on the 20-home module. Primary water distribution is handled by a pair of 6-inch PVC pipes laid in a single trench parallel to the residential street. Average home spacing (lot size) of 100 ft is assumed. With homes on both sides of the street, a 20-home module requires 1000 ft of primary distribution. The secondary water distribution, from the primary to the home water source heat pump, is comprised of a pair of 1.5-inch PVC pipes. Assuming an average home setback of 50 ft from the street centerline, total of 1000 ft of secondary distribution is required per 20-home module. A total of 2000 ft of trenching is required per module. Because this is a new residential community, no disruption of streets or sidewalks is caused by the installation of the water distribution system. In fact, the same trench could be used for the potable water, sewer, electric and telephone utilities. The total 20-home module water supply and distribution system cost of $27,820 is divided equally, yielding a per home capital cost of $1,391.

Table 13.2 illustrates the relatively high capital cost of the water-source heat pump system. While the water-source heat pump is the least expensive home unit, the water supply and distribution systems contribute a substantial proportion of the total system cost.

For the water-source heat pump system to be economically justified, it must offer sufficient savings in energy costs to afford the recovery of the high initial capital and annual maintenance costs. Table 13.4 displays the annual heating and cooling energy consumption for the alternative systems in each of the three cities. The energy consumption data for the water-source heat pump systems includes the energy consumed by the water supply and distribution systems.

TABLE 13.4. Annual Heating and Cooling Energy Consumption

System Alternatives	Electricity (10^3 kWh)		
	BIRMINGHAM	DETROIT	MIAMI
Air-Source Heat Pump	14.69	17.85	11.17
Water-Source Heat Pump (Includes Water Supply and Distribution Systems)	11.66	14.07	9.01

Table 13.5 illustrates the total annual cost computed for the alternative systems in each of the three cities assuming the cost of electricity to be $0.04 per kWh. It is apparent from Table 13.5 that with the base case assumptions the water-source heat pumps or thermal energy storage in aquifer system enjoys a slight economic advantage over the conventional air-source heat pump alternative. Because of the uncertainty associated with the future cost of energy, a graphical technique was developed to measure the impact of changing costs. To make such a graphical comparison, the total annual cost of the two alternative systems is computed as a function of the energy cost. The ratio of these costs is calculated and plotted versus the energy cost.

Figure 13.1 illustrates the impact of variable electricity cost on the relative cost of the air-source heat pump system versus the water-source heat pump system. The cost of electricity is varied from 0 to 20 cents per kWh. The curves plot the

TABLE 13.5. Annual Total System Costs

System Alternatives	BIRMINGHAM	DETROIT	MIAMI
Air-Source Heat Pump	$ 958	$1,085	$818
Water-Source Heat Pump (Includes Water Supply and Distribution Systems)	$ 916	$1,012	$810

Base Case Assumptions:
    Electricity cost $0.04/kWh
    20-home Module
    20-year System Economic Life
    10% interest rate

ratio of the total annual cost of the air-source system to the total annual cost of the water-source system for each of the three cities. Where the curves lie above the break-even line (ratio = 1.00), the water-source system is least costly. On the other hand, where the curves lie below the break-even line, the alternative (air-source heat pump) is superior.

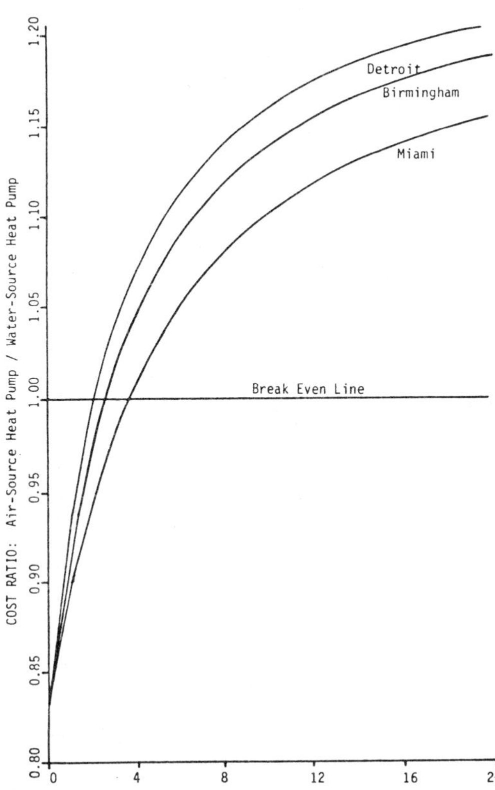

Fig. 13.1. Impact of variable electricity cost.

It is of interest to note that the curves for all cities cross the break-even line at an electricity cost of 2 to 3.7 cents per kWh. These break-even costs are below the minimum cost at which domestic electricity could be obtained in 1980.

Notice that the water-source heat pump system gains its greatest economic advantage in the colder cities (higher heating loads) as electricity costs increase. In no case does the proposed system have a cost advantage ratio greater than 1.20 ($3941/$3263). Notice the relative flatness of the curves above electricity costs of ten cents per kWh.

The graphical technique of sensitivity analysis presented in this case study can also be applied to non-energy cost factors such as the distance between residential units, Fig. 13.2. The base case system assumed a spacing of 100 ft with homes on both sides of the street. The unit spacing is likely to vary widely depending on the type of residential community. A garden apartment or townhouse community might have an effective unit spacing of 25 ft or less. On the other hand, single family homes on large lots, or on only one side of the street, or where all homes in the community do not utilize the system could easily increase the effective unit spacing to several hundred feet.

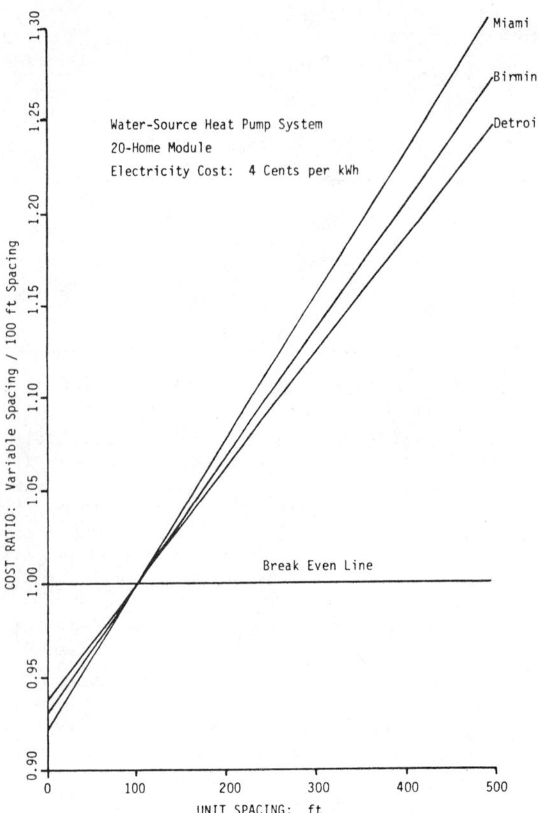

Fig. 13.2. Impact of variable unit spacing.

For this analysis, the base case of the water-source heat pump system is used as the cost ratio denominator. The total annual cost of the proposed system generated by the variable parameter (distance between residential units) is used as the cost ratio numerator. Thus, curves above the breakeven line indicate a more expensive system than the base case. The intent here is to determine the magnitude of the cost impact a change in the base case assumptions has on the economic viability of the water-source heat pump system.

The curves for the various cities are quite close to one another. Note the curves are linear, a more accurate analysis which included the additional energy consumed by the enlarged water distribution system would shift the curves slightly concave upward. The curves indicate that reducing unit spacing to 25 ft would reduce the total annual cost by about 5 percent. Increasing unit spacing to 500 feet would increase the total annual cost by 25 to 30 percent. Such a large cost increase may make the proposed system economically unfeasible in semi-rural residential communities or where less than 100 percent saturation must be expected.

Case II, Urban Commercial District

An urban commercial district was studied for possible retrofit with a system involving thermal energy storage in aquifers (14). The commercial district studied included a set of major buildings. These consist of city-owned, state-owned and privately-owned buildings. Some buildings would be retrofitted and some buildings would have small distributed heat pump units.

Many of the buildings have used water-source heat pumps in the past. A once-through system was used with water being discarded into the sewage system. The heating and cooling systems were converted to standard oil heating and water tower chiller systems for two primary reasons. First the demand for water became excessive and the aquifer water level fell drastically. Second, the treatment of sewage made discarding water economically prohibitive. The proposed system eliminates both these problems by returning most of the water to the aquifer. Some net water discharged is required to keep the aquifer below basement levels.

Excess water can be discarded to balance the annual energy cycle. If excess heat from air conditioning is available, warm water can be discarded or, if the heating load is larger, excess cold water can be discarded.

The capital costs of the proposed system are presented in Tables 13.6 and 13.7. The total capital cost including wells, water distribution and heat pumps is:

Water supply and distribution system	$1,238,390
Heat pumps	2,412,600
Additional piping and installation (20% of heat pump costs)	482,520
Total estimated cost	$4,133,510

The alternative system is in place and thus its capital cost need not be considered.

The additional capital cost of the proposed system involving thermal energy storage in aquifers is expected to be offset by benefits in the form of reduced energy costs. Because the standard system uses both fuel oil ·and electricity it is necessary to convert all energy usages into dollars of cost. For this analysis the prices that prevailed in Fall, 1979 were used:

Electricity: $0.029/kWh
Fuel Oil; $0.70/gal = $4.82/10^6 Btu

TABLE 13.6. Estimated Heat Pump Capital Costs

Units Req'd	Size (Tons)	Cost Unit	Installation Cost	Installed Cost/Unit	Total Cost
210	4	$ 2,300	$ 140	$ 2,440	$512,400
300	5	2,700	150	2,850	855,000
1	100	20,000	1,000	21,000	21,000
2	200	32,000	1,700	33,700	67,400
26	250	35,000	1,800	36,800	956,800
					$2,412,600

TABLE 13.7. Proposed System Capital Costs Not Including Heat Pumps

System Components	Requirement	Installed Costs
Water Supply System		
Wells (120 ft x 12 in) @ $10,200 each	48	$ 489,600
Pumps + Controls @ $10,000 each	48	480,000
Water Distribution System		
Double 12-in Ductile Iron @ $38/ft	6,000	228,000
Gate Valves, 12-in @ $980/each	17	16,660
Trenches		
Excavation and Backfill		
@ $1.48/ft 4 ft wide x 3 ft deep	4,380	6,480
Pavement Patching		
3 in. thick asphalt @ $4.00/ft	4,250	17,000
Curb Replacement		
Concrete 6 in x 12 in @ $25/each	10	250
Sidewalk Repair, Reinforced concrete		
4 in thick @ $40/each	10	400
Total System Capital Cost		$1,238,390

The annual cooling costs were estimated at:

Proposed system; 1.00×10^7 kWh/yr x $0.029 = $290,000/yr
Current system; 1.34×10^7 kWh/yr x $0.029 = 89,000/yr

The annual energy cost savings for cooling equals $99,000/yr. The annual heating costs were estimated at:

Proposed system; 5.9×10^6 kWh/yr x $0.029 = $171,000/yr
Current system; 1.66×10^{11} Btu x $4.82/10^6$ Btu = $800,000/yr

The annual energy cost savings for heating equals $629,000. Thus the total annual energy savings of $99,000 + $629,000 = $728,000 was estimated. The annual cost for system operation and maintenance was assumed to be five percent of the initial capital cost or $207,000 for the proposed system.

The ESIS program (15) was used to simulate the investment options presented by Case 2. The data developed above were used as the base for the analysis. The total investment of $4,134,000 was assumed to be made at the end of year zero. The entire investment was assumed to be borrowed at twelve percent interest with repayment in ten equal annual installments. The system's life was assumed to be 20 years with no salvage value at the end of that period. The annual benefits due to energy savings were $728,000. The annual operating and maintenance cost was $207,000 or five percent of the capital cost. The analysis was performed both with and without income tax. When income tax was included a 48 percent rate was used with a ten percent investment tax credit and sum-of-years digits depreciation over a ten year period, where the IRS allowed extra twenty percent first year depreciation.

The analysis indicated ROI rates of 20.5 and 16.8 percent and payback periods of four and six years for the taxable and non-taxable cases respectively. It is of interest to note that the ROI was greater and the payback period shorter for the case involving taxes than the non-taxable case. Certainly the case in which taxes were paid returned less money to the investor than in the non-tax case. The cash flow timing differed considerably in the two cases. The investment tax credit effectively cut the initial investment by ten percent while the rapid depreciation method provided negative total annual taxes for the first several project years. These net tax gains in the early years more than offset the required taxes in the later project years. Hence the higher ROI and shorter payback period. It should be noted that if the system were owned by a private individual or corporation it might also be subject to property and real estate taxes. Such taxes were not included in this analysis.

The prior analysis assumed a fixed benefits schedule or constant energy prices over the span of the project. Because such a condition is unlikely, a separate analysis was conducted with energy cost (benefits) assumed to rise ten percent per year. Such an increase would be expected to favor the water-source heat pump system with its lower energy consumption. The ESIS analysis did demonstrate a slightly greater ROI for the case with increasing energy cost but the changes were on the order of 0.02 percent and hence insignificant.

A further analysis was made assuming the existence of uncertainty in the estimates of the analysis parameters. The probabilistic data were assumed to follow the normal distribution. The values previously used for the capital investment, annual benefits and annual costs were used as the distribution means with a coefficient of variation of 0.2. The system was assumed to have a mean life of fifteen years with a standard deviation of two. The maximum system life was twenty years as used with the deterministic case.

The impact of the introduction of uncertainty in the analysis was a reduction in the expected ROI from 20.5 to 15.6 percent in the taxable case and from 16.8 to 9.5 in the non-taxable case. The expected payback period was increased from four to 5.1 years and from six to 8.4 years respectively. Considerable variability existed in these estimates with individual ROI's ranging from 0 to 35.7 percent and payback periods from three to fourteen years. Again the introduction of the energy cost trend had a slight impact on the ROI and payback period.

Analysis Uncertainties

Evaluating the economics of a system in its conceptual design stage leaves numerous uncertainties. Fortunately, the proposed thermal energy storage in aquifer systems use existing hardware components with known operating and economic characteristics. Their performance as an integrated system has not been validated, yet looks technically feasible. Economically any new system must be expected to incur a front-end developmental cost. Following a typical learning curve, these costs can be expected to reduce as system design and management experience grows. The analysis presented in the two cases studied does not consider such front-end design costs, but rather looks at potential systems operating after suitable experience has been gained. The base case analysis further assumes the existing component performances to be unchanged.

Rapidly increasing energy costs present a problem in engineering economic analysis. Both Dhavale, et al (6), and Freidenfeld, et al (7), discuss the analysis problems caused by inflation. This problem requires further theoretical study to develop practical rules for application by the engineering economist. System acceptance is another analysis uncertainty. For example, in residential retrofit applications, what if all the possible users do not join the system? Figure 13.2 partially addresses this question from the standpoint of effective unit spacing. But other questions remain related to zoning, financing and disruptions during construction.

References

1. Barish, N.N.,and Kapland, S., _Economic Analysis for Engineering and Managerial Decision Making_, McGraw-Hill, New York, (1978).

2. Bezdek, R.H, Hirsberg, A.S., and Badcock, W.H., Economic Feasibility of Solar Water and Space Heating, _Science_ 203, 1214-1220, (1979).

3. Christian, J.E., _Unitary Water-to-Air Heat Pumps_, Oak Ridge National Laboratory, Oak Ridge, (1977).

4. Clark, E.D., and Lorenzoni, A.B., _Applied Cost Engineering_, Dekker, New York, (1978).

5. DeGarmo, E.P, Canada, J.R., and Sullivan, W.G., _Engineering Economy_, MacMillian, New York, (1979).

6. Dhavale, D.G. and Wilson, H.G., Breakeven Analysis with Inflationary Cost and Prices, _Engineering Economist_ 25, 107-121, (1980).

7. Freidenfelds, J. and Kennedy, M., Price Inflation and Long-Term Present-Worth Studies, _Engineering Economist_ 24, 143-160, (1979).

8. Grant, E.L., Ireson, W.G., and Leavenworth, R.S., _Principles of Engineering Economy_, Ronald Press, New York, (1976).

9. _HUD Residential Solar Economic Performance Model_, Booz, Allen and Hamilton Inc., Bethesda, (1977).

174

10. Means, R.S., *Building Construction Cost Data*, Robert Snow Means, Duxbury, Mass., (1978).

11. Newman, D.C., *Engineering Economics Analysis*, Engineering Press, San Jose, (1977).

12. Page, J.S., *Estimator's Manual of Equipment and Installation Cost*, Gulf, Houston, (1963).

13. Riggs, J.L., *Engineering Economics*, McGraw-Hill, New York, (1977).

14. Schaetzle. W.J., Brett, C.E., and Seppanen, M.S., *Heat Pump Centered Integrated Community Energy Systems*, NTIS No. ANL/CNSV-TM-25 (August, 1979).

15. Seppanen, M.S., *Energy System Investment Simulator*, ESIS, The University of Alabama, (1980).

16. Smith, G.W., *Engineering Economy: Analysis of Capital Expenditures*, Iowa State Univeristy Press, Ames, (1975).

17. Thuesen, H.G., Fabrycky, W.J., and Thuesen, G.J., *Engineering Economy*, Prentice-Hall, Englewood Cliffs, (1971).

18. White, J.A., Agee, M.H., and Case, K.E., *Principles of Engineering Economics Analysis*, Wiley, New York, (1977).

INDEX

INDEX